NATURAL PEST CONTROL

ALTERNATIVES TO CHEMICALS FOR THE HOME AND GARDEN, FARMER AND PROFESSIONAL.

Andrew Lopez
The Invisible Gardener

of Malibu
Since 1972

NATURAL PEST CONTROL

Alternatives to Chemicals
for the Home, Garden,
Farmer and Professional

**The Invisible Gardener
29169 Heathercliff Rd,
Suite 216-408,
Malibu, Ca. 90265
1-800-354-9296**

ISBN 1-885489-07-2

Table of Contents

FOREWORD

It is my pleasure to be able to introduce to you this manual of organic information for the home and garden, farmer and professional, written by an unusually well-informed and qualified person, Andrew Lopez or as he is better known" The Invisible Gardener".

We all know by this time, that our planrt is in trouble, and that we are the ones who must do what is necessary, to save her. Each one of us can work on our small piece of the world, to set up a viable eco-balance system. Understanding the alternatives to chemicals is an important step in this process. It is necessary for us to realize that a change in the way we treat our own environments can make a big difference in the general eco picture.

Chemical fertilizers and pest controls create enormous damage to the environment(poluttion does begin at home!), because they sterize wherever applied(meaning they kill off all the favorable bacteria necessary for a working eco-system, and contaiminate ground waters and kill fauna. Everything that we have learned to do with chemicals can be done with the organic system. The Invisible Gardener gives you all the information you need to do it. Starting with the recipes(formulas), adding his own unique methods, and giving you the sources of whatever you need.

The Organic system is not something to be used only in the vegetable garden, or to control ants, or to avoid toxic products. AS expressed by The Invisible Gardener in this splendid book, it means total commitment to healing the good earth naturally. This often means proper fertilization and foliar feeding so that plants can protect themselves. It also means new concepts for tree management, flower propagation and even compost making-plus special remedies for problems inside the home and out.

The Methodology is all there-how to swtitch away from toxic technology, and how to avoid pollution of air, land and water. I especially like being able to use, say, tobasco sauce to control pests eating my vegetables, or baking soda for my roses, or how to use rock dust on my trees, roses, gardens, and lawns.

Andy's knowledge extents into the house too. One of my favorite chapter is "Dances with Ants". He also saved me a great deal of money in helping me with carpenter ants(I thought they were termites!) and with brown recluse spiders which drop in from time to time.

It's all so easy, safe and economical too!
We know how to make our planet into a Garden of Eden. It is time we do it. Here is a tool. I recommend we all use it. Each one of us should have our own Green Acres.

Yours Organically,

Eddie Albert

CHAPTER 1

Dances with Ants

A bout the Ant

Ants are one of the most familiar of insects in the world. They have smooth bodies, vary in color from black to red, and are divided into head, thorax and abdomen. Visible antennae. It has been said that there are more ants on earth than there are stars in the visible sky(who counts that high?). I have also heard that it takes approximately 500,000 ants to add up to one pound, yet the combined weight of all the ants on earth is greater then the combined weight of all humans on earth. Also Ants outnumber all other terrestrial animals, and live almost anywhere on earth they want. In other words there are a lot of ants out there!

Ants can lift from 6 to 10 times their body weight for a distance of 300 times it's own length even uphill and over obstacles, making them the strongest creatures on the planet (in relation to size). Combine their super strength with their advanced intelligence and advanced social structure, and you will understand why the ants have not changed much in their basic structure in the last few million years, while other creatures were evolving and changing form. It is this near-perfect form that allows ants to live happily in many different types of habitats, from a forest environment to a city environment. These develop into different species each according to their environment. There are so many species of ants with so many interesting habits that I cannot cover them in this book, I would suggest that the library will help here.

A few of the more interesting ants are the thief ants which live in the walls of the larger ant's nest as mice live in the walls of houses; the umbrella ants, which cut leaves and carry them over their heads like

umbrellas as they travel around,; the leaf cutters, which cut leaves, roll them into balls and allow fungus to grow on them. A fun book to read is "The Prince and His Ants" by Luigi Bertelli.

Ants are just as comfortable in an apartment as in a swamp provided there's plenty of food and water. Ants have even been found on board airplanes, on boats in the middle of the ocean, as well as in speeding trains.

While ants do not eat much plant material the Ants themselves are among the main predators of other living things from other insects to animals, birds etc. In Africa, ants are known to bring down an elephant for food, by climbing into its ears and eating it from the inside out. Some ants chew leaves and spit it out in their under ground farms, which is used as food for their underground gardens of fungi. They will raise fungi thru out the year providing their colony with an endless food source.

To watch a single ant hill with it's busy inhabitants moving in some timed motion as if directed by a common mind, gives one some idea, though only a slight one, of the work that is being carried out below. Watching from high above unnoticed, one can not help but wonder of the intelligence behind it. One ant maybe busy dragging a grasshopper much bigger then its self, while another ant is following a scent trail that will lead it to a food or water source. He goes from sign post to sign post, remarking the trail for others to read and follow. They know exactly where your kitchen is!

Ants are clean insects and are constantly cleaning themselves with their antennae.They have developed a very sophisticated system of communication. When ever their colony needs anything the call goes out and the workers tell the scouts what is needed and off they go looking for just that food. When they find it they relay the message back to the colony and soon every one in the colony will know it, and off go the workers in force to get it and bring it back.

One of the most important changes in the ants have been in their behavior. Ants have learned to survive and recover very quickly from any disaster, especially that of man's doing . They have adapted to benefit from mankind. They have quickly learned how to train us. They know how we will react to any given situation and know what buttons to push. They have adopted us as they have adopted the aphids! They remember events for several months. Leave a piece of food or sweets to be found by them and they will have your place mapped and marked with special paths that lead right to it. It is for this reason you will not be able to eliminate ants entirely from your environment, and that you would be wise not to do battle with them, but to communicate with them and to reach a mutual understanding.

More Ants

Ants belong to the hymenoptera family of insects (like bees and wasps), and to the formicidae family, which means simply that they live in colonies. In each colony there are three classes of individuals: The females or queens (fertile females), males (used for mating only), and the workers (wingless sterile females). The workers are the busy members of the family, they are also sterile females but are highly endowed with brain power and and mechanical skills. Among them are three classes: The main workers, minor workers(gardeners) and the soldiers. The first two differ in size; the soldiers have huge heads and powerful jaws. They are seen guarding the nest, defending the nest against attacks from other ants, etc. Worker ants can live from one to four years while some queens live up to 20 years. The home of the ants differ greatly depending upon the kind of ant and country(environment) it has to deal with.

There are several stages the ants go through in establishing their colony, which is important to know in order to discourage them from moving into your household.

In the Beginning

Before they swarm, the males and queens are the aristocratic members of the family and are treated as such. They are cleaned and fed well. Their heaviest labor is mock battles or a game of tag: your it!

It's Party Time!

The first stage starts with the nuptial flight when the virgin females leave the colony of their parents and take with them some of the males(both are winged). They explore the surrounding neighborhood and perhaps some distant suburbs looking for a nice place to start a new home. There are far more females than males. Many die within the hour after being released as they fall prey to birds, spiders and other predators. The females fly into swarms of males and mate, often with more then one male in mid air! After mating, the inseminated females search for new locations for their colony to start, while the males (not being needed and serving no useful purpose) either die at the hands of their lover (the queen) or if lucky, may manage to leave the queen and survive for up to three days.

The Queen has to either rejoin her old colony, join another colony or start a new colony, as mentioned before, or the queen can also attack another colony and kill the current queen, taking over her colony. Once she has found her new location, she will drop off her wings and dig herself a nest and seal herself in it to lay her eggs.

Depending on the species of ant, the queen will either seal herself off permanently and will stay in total isolation (called claustral), or she will forage outside for food (called partially claustral). There are two types of swarming. Budding, when workers leave the main colony with one or more queens and start a new nest, and fission, where parts of the colony containing the fertile queens separate, taking with her a large portion of workers.

The Ergonomic Stage

The next stage of colony growth is called the ergonomic stage where the first workers born develop into mature workers. These first workers are usually small due to shortages of food and will get larger with each generation as more food becomes available to the colony. The main purpose of this stage is for colony growth. Here the queen will lay only fertilized eggs, which become females. Whether they become workers, soldiers or new queens depends on the care they will get and upon the chemical signals received from their nurses.

NATURAL PEST CONTROL

The Reproductive Stage

The next stage is the reproductive stage, when the colony begins to produce males and fertile females. This is where the queen produces unfertilized eggs that mature into males and at the same time, the workers will begin to groom female eggs into new queens. When they have matured, it is time for the cycle to begin again with the nuptial flight.

Ants have very advanced societies, complex and highly organized. Some ants live on household foods, sweets or protein, some eat only aphid honeydew, some eat seeds, grains or vegetable roots, some grow their own food such as fungi, and of course most ants eat other insects.

"Ants are involved in more than 70 percent of problems associated with both house and garden plants."

Ants help to spread various different types of bacterial diseases; some ants help spread the virus through their 'farming' practices of herding aphids from one part of the plant to another. Ants will occasionally run into a group of aphids. To the ants these aphids are a great prize. The ants will stroke the aphids with their antennae until the aphid secretes a drop of a sweet liquid known as honeydew; this the ant stores in his second stomach. He does this until his second stomach is full only then will he ingest honeydew for himself which goes into his first stomach. The honeydew taken back to his nest is taken to the nursery where the ant babies or any one else in the nest who needs to be are fed. The nectar is, unfortunately, not cleaned up very well after a meal, and the left over nectar then becomes food for many different types of bacteria and viruses. It also attracts many other insects such as whiteflies, scales, mites, and so on.

Thru out the years ants have learned that aphids are valuable food producers not to be left alone as the prey of other tribes that might also find them, so they start to keep the aphids nearby inside a special corral built for them by their masters deep in their tunnels. The aphids are happy slaves and will not only produce food for them but help in digging out the tunnels. The ants will also care for the aphids baby's as they do for their own. When they need more aphids for the herd they can go out and hunt the wild aphids since not all aphids are domesticated. Also not all aphids have this relationship with the ants.

"If you can control the ants, you will have a greater chance of controlling your general insect problems and plant diseases, provided you maintain a healthy, balanced environment. Spraying a chemical pesticide, even one that rids you of your ant population for few weeks or even months, does not solve your ant problem."

Do Not Walk on the Grass

If you employed one of the ubiquitous services that claims to solve the ant problem with chemicals, every few months someone would come by and spray your property, put up the now famous "Do not walk on the grass for 48 hours" sign, and that would be that.

There are many problems that arise from employing this method of pest control. The land, groundwater and produce are all polluted. Numerous health hazards arise that result from exposure to these chemicals. And, most importantly, this method only increases the ants' resistance. Ants quickly develop an immunity to all chemicals, thus requiring stronger chemicals the next time around.

The No-Pollution Solution

The solution to these problems is straightforward: Stop using chemicals and use a different approach. Viable alternatives exist to chemical fertilizers, pesticides and herbicides that kill organisms within the soil, and upset its natural balance.

NATURAL PEST CONTROL

"There are two ways to sterilize the soil: by using heat and by using chemicals"

Chemical fertilizers completely sterilize the soil, killing off the beneficial organisms (called simply "beneficials") that live in it. Predators then gain the upper hand while fewer beneficials are able to re-establish themselves. When this natural balance is upset, the natural systems break down, causing infestations of one pest or another.

Nature's Janitors

Ants play a very important role in the plant kingdom. They are not pests, but nature's janitors. When kept within their boundaries, they perform many functions which are important not only to the plant kingdom, but by extension to all living things. Ants reveal to us the importance of nature's balance. They serve as indicators that tell us when there is balance and when there is not.

It is unnatural and impossible to eliminate ants from the face of the earth. Remove them and a great deal of damage will occur. The answer lies in learning to control them. To keep ants off your roses, out of the house, and off the fruit trees, you will have to understand the various relationships that ants have with various plants, animals and humankind.

Timing is important. Ants help pollinate fruit trees and flowers. Thus, they must be allowed on the plants during this time for proper pollination, but not at others times, such as when the fruit trees are bearing.

It is during this time that you use barriers to keep them off. One of the best ways to deal with the ant problem is to realize that properly-fed and healthy plants are less attractive to ants, other insects and diseases than plants that are sick, weak and under stress due to improper feeding and watering.

"There is a relationship between nutrition and pest activity"

Providing for your plants only 100 percent organic natural fertilizers is a very important step in reducing plant stress. Using compost and natural fertilizers helps keep the soil alive. It is this life that provides plants with the nutrition they need to keep insects at bay. The greater the nutritional levels, the less the stress levels.

Use liquid seaweed to spray the leaves. It is best to blend the different types of liquid seaweed to obtain a complete collection of all the trace minerals needed for healthy plant growth. Water your plants regularly. Set up a pattern of watering and stick to it. If you are in an area where conservation is required, use drip systems.

Mulching

Mulching is very important as ants love dry, unfertile soil. The importance of compost cannot be overstated. Compost properly made will provide an important source of nutrients and bacteria, needed by the plants for proper growth.

A Few Hints

Ants have built in radar for finding food, water, and whatever they need. They know when plants are stressed and act accordingly. They have been around much longer then we have. The land your house is on was inhabited by ants long before you. The earth is their home as well.

Super Highways

Ants build what could be described as underground condominiums, and use 'highways' to get from one place to another. They don't have to build these superhighways, since they prefer using paths that humans, other animals, and other insects have already established. They choose the path of least resistance, following sprinkler lines, drip lines, sidewalks, hose lines, or any other route that gets them where they are going with the least amount of effort.

NATURAL PEST CONTROL

Follow the Yellow Brick Road

When using natural sprays and barriers (which is described in a later chapter), you must first find these paths, then spray or place the barriers so the ants will be routed away from where you do not want the ants to be. You will find that it helps to provide food and water. This should be done in an area where you don't mind their presence. Remember, you will not be able to completely rid your property of ants. Even if you did, new ants would simply move in before you know it.

The key to dealing with ants successfully is a balanced environment. Nature plays by certain rules. Respect these rules and they will respect you. Instead of imposing your own rules on nature, learn how to achieve a balanced environment within your own ecosystem.

Nature will naturally keep ants in line, provided you help her out. Since we are part of nature, you might say she's helping herself. Give her time to regain her balance. One year should suffice, but the important thing to remember is that it will take time to control ants. It will be time well spent and the results last forever.

Manufactured Demand

We do more damage to our environment than ants ever could. The chemicals we use to kill ants will never work with any real success, because ants eventually return stronger then ever. Many of these chemicals are left-overs from the stockpile of chemical warfare weapons developed during WW II. Newer, more deadly strains of these substances are developed every year to meet the demand of chemical farmers.

Chemicals applied directly to the soil will purge it of all living things. Once sterilized in this manner, soil will not support life. Dead soil is a nonfunctioning system. Plants growing in it will experience great stress and eventually die.

"It is that simple, To Heal the Earth and help Nature to regain her strength, we must stop using any chemicals that cause imbalance"

Here are the Seven steps that will help you to start controlling the ants around your home and garden. The steps are simple and easy to do.

Steps in Controlling Ants
Step 1 Buying your own Ant Cafe

The Ant Cafe(TM)
The Ant Control Center

This ant control center is a simple Finch bird box with a secured lid. It has a small hole in the front through which the ants enter. This hole must be small enough only for ants 1/4 inch is enough. The box can be bought already made for small baby birds to feed from(called a finch box see drawing above). The cost is about $5 and is available at any pet shop. It's worth the cost. They will last you several years.

Place an eight-ounce plastic cup inside the box. These are handy for putting in the ant gel. Make sure that the lid is screwed on so that small kids and other animals (dogs, cats) won't be able to get at it either. Remember to hose out the Ant Cafe every month and throw any left over ant gel into the trash.

Do not place down yet. That's step 5.

NATURAL PEST CONTROL

Step 2....

If any ants are coming into the house or any other place(s) where you do not want them to be, 1st follow their trail from within the house. See where they are getting into the house. 2nd Follow the trails along the outside of the house and place an Ant Cafe in the area, making sure the ants know it is there by placing honey on their trails leading to the Ant Cafe.

Step 3....

Buy some Dr. Bronners Peppermint soap. Add a few drops into a quart sprayer filled with water. Start by spraying the kitchen then the bath room. Try to work your way from the center of the house out towards the edges. Spray lightly in the areas the ants have been seen in. Don't forget the bathroom.

Tell the ants the day before you do this and that you want to make a deal with them. Tell Them that you are willing to feed them outside your house and that you don't want them in the house. Also that you are going to spray the Dr. Bronners to make sure they leave the house and also to confuse the ants by hiding their ant trails.

Step 4....

Once this is done, then caulk up the entrances and exits that you found the ants were using to get into your house. Sometimes this is not possible to find.

Step 5.....   VERY IMPORTANT!!

Put honey and water into the Ant Cafe cup and screw the lid down and place where ants are seen and where the Ant Cafe can be secured down. This is usually done along the outside of the house. Usually 4 Ant Cafes are placed on the ground along the outside wall. Cover with large rocks so only the ants can get at it. Keep out of direct sunlight as it will ferment it. The ants will enjoy this free meal and will consider it a food haven. This is important to establish. Do this for the first month or so, keeping the Ant Cafe cup filled with honey water or sweetened water. This may solve the problem by itself and you won't have to use a poison(boric acid).

Gradually move the box away from the house. You will notice that the ants have stopped going into the house and are going into the Ant Cafe.

Providing them with a food source will be enough to control and keep them out of your home and garden. This is rather simple, but maybe enough for your needs. In this simple case, a cup or bowl will be enough, simply have sweetened water available to them, placed along their trails. Use barriers to keep them out of the house (explained later). Ants will find food with or without your help. Here is your chance to modify their behavior.

Step 6....

Continue to spray the Peppermint soap as often as needed inside the house while checking the ant cafes at least once per week to make sure the ants are using it, cleaning it if necessary. A fungus may grow in the sugar mixture since ants carry their own fungi on their bodies. Just clean it out by disposing of the used liquid in the trash. Replace with a clean solution.

Step 7...

Look around your property, ask yourself if you are maintaining a healthy environment. Composting and mulching, controlling water, these are the tools you must use in keeping the ants under control naturally.

A.I.P.M
(Advance Integral Pest Management)

Developed by The Invisible Gardener. IPM states that if you have exhausted all organic alternatives and you still have a problem with pests, then certain chemicals could be used. AIPM states that if after following steps 1 through 10(of the IPM manual) you still have pests then go back to step 1!

NATURAL PEST CONTROL

Here is the formula for the liquid Ant Control mixture which you use inside the Ant Cafe

Liquid Ant Control Gel #1
7 oz water
white sugar or any sugar source
2 oz Honey

Use a different source of sugar each month. One month use honey, next month use white sugar cubes, the following month use corn syrup, etc. Ants are smart and learn very fast and they also remember what they have learned! Place the 8 oz plastic cup inside the Ant Cafe. Blend the 2 tablespoons honey in the 7 oz water, add a little white sugar, to sweeten.

Liquid Ant Gel #2

This is a protein version of above, instead of sugar use protein sources like cat or dog food or peanut butter. Pour the ant gel mixture down into the small plastic containers in the Ant Cafe and allow to over flow so that the ants will find it. It is very important that you show the ants where this feeding station is located by placing pure honey outside the Ant Cafe as well as across their paths. Ants follow scents but very often have lost this ability and therefore must be shown where the food source is. Once their scouts have located this feeding station, they will always come here first.

This station has to be refilled every month (or as often as needed). Close the lid and screw it on with screws. Screw down the Ant Cafe near known ant trails around the house and in the garden, place under fruit trees (hidden under rocks), etc, always out of direct sunlight. Keep in a cool dark place. Heat will cause it to ferment, if this happens wash out with hose and refill with new gel. It is for this reason that you should make only enough ant gel for one use at a time.

Use regular honey and place a few drops directly outside the Ant Cafe , to let the ants know its there. Remember to change the recipe every time you make it. You can place the Ant Cafe inside the house in places like the kitchen (hidden where no kids or cats can reach them), they can also be placed in the wash area behind the washer, and they can be placed in the garage near the garbage cans to control ants there. But this is only advisable in the beginning and they should be brought out to the property ASAP. Make sure you screw down the lid so that no kids, dogs, etc, can get at it.

Dancing with the Ants

Learning how to make the Ant Cafe an active food source for the ants is very important. It can not be overstated. The ants will not be able at first to find this food source for many different reasons and sometimes they will not want it and ignore it. You must make them understand that this is a food source for them. This is what we call Dancing with the ants. Therefore it is good to go slowly at first when introducing the Ant Cafe to the ants.

Providing food and water for the ants while using the peppermint soap inside the house maybe enough to control your ant problem inside the house.

We call this part Dancing with the ants because it takes time to get the ants to go for the ant cafes and not go into your house or on your roses.

NOTE ABOUT USING BORIC ACID

Sometimes it is not possible to control the ants by feeding them. **When the imbalance is great enough there is infestation.** An infestation is indicative of a greater problem. Correct that and you will be better able to correct the ant problem. However you will have to reduce the ant population in order to regain partial balance. The addition of boric acid to the ant gel will do just that. The ants will find it and take it back to their colonies and sooner or later everyone will partake of it, killing them and since it won't kill them all, they start over again.

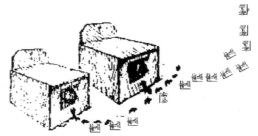

You must make the boric acid just strong enough to slowly kill off the colony. If it is too strong it will kill the scouts right away before they can make it back to the colony. Use 1/4 tablespoon per cup.

Even tho Boric acid is sold as an eye wash, Boric acid is dangerous when taken internally. avoid cuts). So please be very careful using . The mixture should be a thick fluid(like honey). You should make just enough for your immediate use. Boric acid can also be bought as 20 mule team borax soap.

Boric acid, when used in concentrated amounts, is dangerous to animals, bees, cats, dogs, small creatures, trees and humans so be careful not to use outside of the Ant Cafe where children can get at it. Keep out of reach of children. Do not spill around plants or trees, if so water well. If you suspect that your child or animal has gotten into the Ant Cafe, get to a doctor ASAP!!(the amount of boric acid used in this formula is deliberately very low, this is to insure that if accidental ingestion should occur, the mixture would not be strong enough to kill). But never the less do not leave any amount of gel made sitting around. Drink milk and induce vomiting Kill only as a last resort!

Always follow instructions on labels of products that you buy.

Never pour ant gel into any type of drinking container, soda bottle or in any thing that would confuse someone into thinking that it was safe to drink. Always label the ant mixture as poison and write keep out!

Before you add boric acid, ask yourself is it necessary to kill them or was the sugar alone keeping them out of the house?

About using DRAX

Drax is made up of 5% boric acid and 95% apple mint. The ant gel or Drax formula should be adjusted so that the level of boric acid that the ants get is strong enough to kill the queen but be weak enough not to kill the worker who brings it to her. A good way to determine this is to first provide a 100% pure sugar source then after the proper length of time(the ants are busiest and have settled in) slowly switch to

boric acid by first adding a few tablespoon worth of the boric acid into the formula, then see if the ants are dyeing at the site; if so then it is too strong.

If the formula for the ant gel is killing too many ants too fast and the ants are no longer going there, then use 1/2 the formula for boric acid. The Ant Cafes are a very important tool in your pest control program. They are useful for helping to control many other insects as well.

Remember, please use the boric acid only as a last resort and don't use it all the time. I suggest getting 4 Ant Cafes per home and nail around house; 4 per garden, nail to raised beds, and place around fruit trees being bothered by ants. Ant cafes will provide continuous control of ants, carpenter ants, earwigs, and roaches(see chapter on roaches). Many other insects are also attracted to this mixture.

You can make your own ant gel or you can buy Drax ant gel from Peaceful Valley, Natural Gardening Research Center(Gardens Alive!), see resources directory for more.

A note about using Drax or any other manufactured ant gel. These gels work fine for one or two months then the ants will get used to it and leave it alone, that is why it is best to blend Drax with your homemade gel.

CONCERNING THE CONTAINER USED TO BUILD YOUR ANT CAFE

I have experimented with many different types of containers in order to find out which one worked the best. There are many problems with using plastic containers due to their pollution abilities(damage to our OZONE layer from their manufacturing process, damage to our oceanes, etc.) so I do not recommend using this material unless it is made properly.

Here are some important points to keep in mind when you are looking for that' perfect' Ant Cafe:

It is very important that the unit must be able to either be locked, screwed down or in some way made impossible for children, dogs, cats, etc, to get into. We are after only ants here!(if you are us-

NATURAL PEST CONTROL

ing the boric acid) If you are concerned about kids getting in to it then don't use any boric acid and use only sugar.

The container must hold enough ant gel(or plain sugar water) to last a week or two. Be small enough to be hidden from view. Cost consideration. Stay away from plastics if possible or reuse the plastic cups.

ORGANIC ANT CONTROL SPRAYS

Here is a list of some formulas for making organic sprays, what they are, how to use them and some sources.

Any Organic Bio-degradable soap(see chart on safe soaps to use), Scented mixtures(such as mint extract, garlic oil extracts, or tabasco sauce can be used to spray on ants. You can also use scented soaps such as Dr. Bronners Peppermint soap or any natural concentrate made without chemical additives.

Solar Tea Made from any of the mint family such as peppermint(this is tea made by allowing the tea bags to sit inside a glass container filled with clean water and allowed to sit for 24 hrs in the sun).

The Ant Spray Formula for outside only

Add 1 oz Bio-degradable soap (Dr. Bronners or Shaklees Basic H or Amways LOC, or any soap made naturally without chemicals or additives) per gallon of water. You can add either 5 drops garlic oil, or add 2 drops mint extract or 5 drops tabasco sauce per gallon.

Always check the ingredients to be sure that no chemical additives are in it. It is almost impossible to find pure foods so go to a good health food store to buy your materials, or make your own. Try buying your materials from mail order suppliers such as PEACEFUL VALLEY and AR-BICO or NITRON or WORMS WAY., Gardeners Supply, and many other mail order sources.

You will need a good quart Sprayer(for a three gallon unit, simply triple the amounts). The concentration will vary according to the insect you want to kill(experiment with the strength, Add more soap,garlic, tabasco or mint if needed.

The amount that works best for you and for your situation will be different for someone else in a different part of the world, having to deal with a different variety of ant(or any other bugs for that matter).

You may want to experiment with the mixture yourself to see what works best. Use this spray as often as needed; usually once per week will do. Spray around the outside of the house and any places where you see ant

trails or where they are entering the house. Spray only on the ants, flood their ant colonies with it(usually under rocks) and on traveled paths, around the base of your house and other entrances, but DO NOT SPRAY DIRECTLY ON PLANTS AS THIS MIXTURE WILL BURN THEM!, if this happens wash off with water. You can omit the garlic if the garlic scent is unpleasant to you. Avoid spraying this mixture on walls as the tabasco sauce will stain. You can use just The Peppermint soap and water when spraying on walls.

When ants are coming into your house and you want them to stay out. First thing you do is set up the Ant Cafe outside(see Ant Cafe). Then the second thing you do is follow the ants and see where they are coming in, spray inside the house with Dr. Bronners Peppermint soap to make them leave then you plug up the entrances where they are coming in to keep them out.

Then you spray a mixture of Dr. Bronners Peppermint soap and water along the outside of the house and directly on ants. Use 1 capful soap per quart. Test on the ants for strength. If they don't die right away then add another capful and try again. Make stronger as needed. Shaklee's Basic H is another great soap to use. This one has no smell for those of you who don't want any smell! You can also try citrus soaps, herbal soaps, or any safe natural soap.

NATURAL PEST CONTROL

"Something to remember is that you really don't want to kill everything in sight. The idea is to learn to live together in harmony and killing should only be a last resort."

Citronella Oil comes from the Citronella Herb and is an excellent oil to use. Try smelling it first to see if you like the scent (not every one does). You can add a few drops per gallon to the above formula to increase effectiveness. Tea tree oil would be great to add also. See more info on this product in a later chapter on pest control.

Safe Ant Dust

Pyrethrum powder and Dia-Earth make a great mixture in controlling ants. Use 1 lb Dia-earth and 5 oz pyrethrum. Mix well. Avoid breathing. Can be used as a dust where ants are found.Use outside around plants and on ant trails. Use inside under cupboards, sprinkle around windows, cracks. Powdered Pyrethrum is available from AR-BICO, EcoSafe Laboratories, Peaceful Valley, Garden-Ville, Gardens Alive!, Nitron Inc sells a pyrethrum dust and dia-earth mix called Diacide. See Resource Directory for address..

It is 100% safe to mammals, and birds.

There are many products on the market which use liquid pyrethrum. Pyrethrum by itself is an excellent safe pesticide to use against a variety of insects. Pure Pyrethrum powder is best to use here as it does not have any additives.

One of the problems with using liquid pyrethrum products is that a product called piperonyl butoxide is added as a booster. Pure pyrethrum is safe to use and handle. Do not breathe and get into eyes, if you do wash with water.

WARNING!!

Piperonyl Butoxide has been associated with liver disorders. Avoid breathing and wear plastic gloves when handling. Write to what ever company sells this product and tell them that you want them to stop adding it to the pyrethrum. Better yet, tell the manager where you bought it .

At present all liquid pyrethrum products have this additive except Gardens Alive! which sells a Liquid Rotenone-Pyrethrum mixture. I therefore do not suggest using this product (with PB). They also carry Safer Insecticidal soap with Pyrethrum which contains no Piperonyl butoxide.. I strongly suggest you make your own liquid pyrethrum. This is easily enough done.

Making a Pyrethrum Paste

Pyrethrum powder does not blend with water well. You have to first add a small amount of alcohol to warm water and then the pyrethrum to dissolve and form a slurry. The strength will vary according to the insects you are attacking. For more information on Pyrethrum see later chapter on making your own pest controls.

NATURAL PEST CONTROL

A NOTE ABOUT THE WARNING LABEL ON THIS PRODUCT

The EPA classifies pesticides into four toxicity categories, but they use only three human-hazard warning signal words for pesticide labels: Danger(Poison), Warning and Caution. Therefore Natural products such as Bacillus thuringiensis(BT), Garlic, Safer Soap, Pyrethrum, Dia-Earth, etc carry the same signal word as Sevin, Trimec, Dursban, Diazinon, and many other chemicals.

These natural substances cannot be categorized with these chemicals, yet the EPA has not created a fourth signal word. Until this happens, the consumer, if not informed, will pick up Sevin, etc, thinking that these products are as safe as BT,etc. He thinks he has only to use caution.

A Safe Ant Spray

1/2 oz powdered pyrethrum paste
1 cup Dia-earth paste
1 gallon water
1 oz Any Bio-degradable soap
1 tablespoon tabasco Sauce
5 drops Sesame Seed Oil
(Sesame seed oil contains sesamin, a safer and more powerful synergist for pyrethrum then PB(Piperonyl Butoxide).

During WW2, the Armed Forces used more than 40 million gallons of aerosol bombs containing pyrethrum, liquefied gas, and sesame oil. Up to as much as 8% of the formula was sesame oil. Sesame oil is also an excellent synergist for Rotenone.

In a bowl, add 1/2 oz of the pyrethrum paste, 1 cup DE paste, 1 oz of the soap, 5 drops sesame oil, a tablespoonful of tabasco sauce. See how to make a paste in the Preps section. Slowly add enough water to the mixture to make a slurry. Make sure the mixture is completely dissolved. Pour this mixture into a gallon container filled with water, while stirring regularly to aid in dissolving the mixture. Allow to settle then pour through a strainer into a quart or gallon sprayer. To be sprayed directly on the ants.

This formula can also be sprayed directly on most plants without any damage occurring. A test should be made first to insure that no damage occurs to the plant(s) sprayed.

This IPM formula is made up of Diatomaceous Earth, Pyrethrum Dust, North Atlantic Kelp, Rock dust, a dash of Cayenne Pepper, and Sulfur Dust. Can be used as a dust, painted around the trunk, as a paste, or made into a liquid and sprayed.

Plant Guardian Organic Paste

1 lb	Dia-earth(not pool grade)
1 lb	Rock Dust
1 lb	Kelp Dust(or seaweed powder)
1 lb	Alfalfa Meal
1/4 lb	Pyrethrum flowers,Dust
1/4 lb	Cayenne Pepper
5 oz	Sulfur Dust

Mix well together. Avoid breathing or contact with eyes. Dried Peppermint can be substituted for pyrethrium. Remember, the pyrethrium will kill, the pepermint will repell.

Use only a small amount for best results.

Use the Pest Pistol(tm) or Dustin Mizer(available at Gardens Alive, Nitron) to dust plants being attacked by pests, or you can make a spray by mixing 1 cup paste with enough water to dissolve then strain into 1 gallon water. Add 1-5 tablespoons of tabasco Sauce or Cyclone Cider. Can also be made into a paste by slowly adding water until it starts to be

NATURAL PEST CONTROL

workable like clay. Add 5 tablespoons either tabasco sauce or Cyclone Cider. Can be painted around trunks of tree for protection.

Effective against snails, slugs, ants, earwigs, spiders and a wide variety of bugs. 100% Organic. Use the Pest pistol(tm) to form a line around the tree trunk or plants that are being attacked by ants. Repeat as often as needed.

A Liquid Plant Guardian(tm) Formula

The same formula is followed except you add 1/4 lb of paste to 1 gallon of warm water, adding slowly while stirring. Allow to sit for one hour and cool. Then strain the mixture through a fine cloth. Use 5 to 20 drops per gallon for most bugs. You will have to experiment and find out the best amounts to use for your own areas. Use either Citronella Oil(at 5 drops per gallon) or if you don't like Citronella Oil, you can use tabasco Sauce(1-5 tablespoon per gallon).   You can also use Cyclone Cider at 5 drops per gallon. WOW! Controls, kills or repells a wide variety of bugs on contact.   100% Organically safe.

NATURAL PEST CONTROL

Chapter 2

Snail Tales
How to Control Snails Organically

Most problems with snails can be explained by the law of cause and reaction. The snail problem that many gardeners encounter is a reaction and not a cause. Snail infestation occurs when the soil is no longer alive because the organisms within it have been killed with chemical fertilizers, pesticides or thru otherwise bad soil care. The continued use of snail killer year after year destroys the soil and upsets the balance. This is the cause. Dead soil will not support life, therefore the delicate balance which nature maintains has been upset. It is this balance that protects all living beings on our planet. Deal with the cause and the effect will disappear. Here are some steps you can take to control snails on your property.

Step 1: Dealing with Snails

Whenever there is an infestation of a pest, the balance of nature has been upset. Anything not naturally found in nature contributes to this imbalance, including chemical fertilizers, pesticides, herbicides, weed killers, snail killers or any other poison and/or synthetic toxin.

If you have been using any form of chemical snail control year after year to keep snails away from your flowers, vegetables, etc., the first and most important step is to stop using any form of chemical snail bait whatsoever. Another reason to stop using snail poison is that it also kills birds and other friendly critters that you may like having around your home. They are here for the snails to.

If you want to deal with snails effectively, you will succeed only if you understand that chemical fertilizers are as bad for your soil's health as pesticides. You must be willing to commit yourself to growing without chemicals of any kind. Because your plants need a period of transition, start by reducing your chemical use by half during the first 3 months, then completely stop the chemical use after that period.

Step 2: Withdrawal

Allow yourself and your property time to go through this withdrawal period. This is just as important to you as to your property.

"The day you stopped using chemicals is the day you started to regain the eco-balance of your property."

Chemicals only cause imbalance. This withdrawal period is a critical time for you and your property. Without the snail bait, the snails will seem to increase (that's because they actually are increasing at first), you must therefore employ commando tactics.

Step 3: Handpicking

Handpicking snails/slugs is one of the fastest and safest ways to reduce their population. Snails are best picked after dark or on cloudy days when they are most active.

Water well a few hours before picking to promote activity on their part. Throw a snail picking party. Tell your kids that you will pay them per snail picked. A good price is a penny per snail (100 snails = 1 dollar). Have plenty of flashlights available. Provide gloves. Have buckets filled either with salt and soap water, DE salt layer, or rock dust layer. Throw picked snails into a bucket for disposal later.

Alternatively, you can crush the snails and place them into the ant Cfe as ants love to eat snails. If using rock dust, crushed snails can be added to compost pile.

If you have chickens, feed the snails to them. Pick up every single snail you see and destroy it. Snails can produce over 300 eggs per day during laying season and the eggs can stay buried in the ground for up to 11 years and emerge when the time is right for their survival. Another method is to attract and trap them. (See also The Snail Cfe).

Step 4: Restoring the Balance

Increasing the energy level of the soil increases the health of the soil. This process naturally reduces pests and diseases. The best way to increase the energy level of the soil is through proper organic nutrition. Compost that is rich in minerals and bacteria naturally increases the energy level of the soil. The addition of rock dust also increases the energy level of the soil. Higher energy levels support a more beneficial diversity of life.

Beneficial diversity is described as the various interrelationships between living organisms and their contributions to the whole of nature.

NATURAL PEST CONTROL

Step 5: Natural Snail Control Methods

Attracting and trapping snails is one of the most effective natural snail controls. The best way that I have found is through using the Snail Cafe. In my search for a more effective way to deal with snails, several things happened. I found that people generally use one chemical or another to rid themselves of snails. They continue dumping chemicals into the soil even though their snail problem persists. Those people who do not have snail problems do not lace the ground with anti-snail chemicals, concentrating on nutrition and healthy soil, instead. During the transition period when you are creating a healthier environment around your home, a good way to get rid of the snails is through the use of a Snail Cafe.

The Snail Cafe(tm)

Main Unit

The cost to build a Snail Cafe is minimal. It should be made from a natural product such as wood. Wooden bird houses make excellent Snail Cafe. Go to any bird or pet shop and look at their bird houses. A good unit must be big enough to hold a bowl and also to allow snails to enter easily.

Pick one that allows enough room for the snails to enter, and one that has a lid that can be opened and closed. This prevents the snail bait or snail beer from getting into the soil and insures that you only kill snails (if using boric acid). This unit will also protect the snail bait/Brew from rain or sprinklers.

Inside of the cafe you place your homemade organic snail bait or

snail beer. You should alternate between using snail bait and snail beer. This will increase the effectiveness of your Snail Cafe. Place in a shady spot. Remove snails everyday, weekly or as often as needed.

Discard the snails into the compost pile or use in the Ant Cafe. Replace with new snail bait or snail beer. Use as many Snail Cafe as needed.

The Snail Bait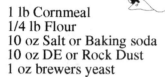

1 lb Cornmeal
1/4 lb Flour
10 oz Salt or Baking soda
10 oz DE or Rock Dust
1 oz brewers yeast

Mix all ingredients together and place into snail hotel. Feeding the snails with this mixture will dehydrate them. Be patient as this takes time. Boric acid can be used at 1 oz for the above mixture but use only as a last resort and if snail infestation is great.

The Snail Brew (To attract them by the thousands!)

1 Quart Glass bottle
1 tablespoon Brewers Yeast Apple Cider Cheap Wine or Beer
1 tablespoon natural soap
1 cup warm sugared water (add 1 tablespoonful of sugar or honey).

In one-quart bottle, add one cup cheap wine or beer and 1 cup apple cider. Mix with one tablespoon of brewers yeast and enough warm honey water to fill. Add one tablespoon of any natural soap such as shaklees Basic H or Citrus Soap (see natural soap chart).

Do not shake but stir well. Pour into plastic cup or bowl and place into the cafe (using the apple cider, wine or beer to control snails works because of its high fermentation).

Make sure the lid is open to allow the snails to enter. Place near their favorite hiding places. You will have to check it regularly, to remove any dead snails and or replace with new snail brew.

Providing snails with hiding places makes them easier to locate and dispose of. Old nursery containers placed on their sides make excellent temporary hiding places for snails. You can place a bowl of snail brew inside these containers to attract them. You can also use boards laid on a rock to allow snails to crawl under. Pick them off regularly.

Shoot'em Up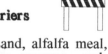

If you do not want to hand-pick the snails you can always try the Cowboy School of Snail Eradication. Prepare your amunition by mixing one-half salt with one-half cayenne pepper into a salt shaker. Carry the shaker with you and sprinkle the contents on any snails or slugs you encounter. This causes them to dehydrate and die. Another method is to use a spray bottle with a mixture of one-half ammonia, two tablespoons of any natural soap and remainder with water. If you don't like ammonia, you can mix one-half vinegar, two tablespoons soap, and two tablespoons tabasco sauce. Spray either of these two mixtures on the snails to kill them. A third mixture is to use tobasco sauce(about 10 drops per quart water) and spray directly on the snails.

Snail Barriers

Silica sand, alfalfa meal, rock dust, flour cayenne, and seed weed all make excellent barriers to keep snails from crossing over into an area you want to keep them out of. Use as little as needed around plants to protect them from snails. Tangelfoot made of castor oil and wax and placed around the trunk of a tree also makes an excellent barrier. Add cayenne pepper to increase effectiveness. A foot wide barrier

NATURAL PEST CONTROL

should be placed around the edge of your garden or beds. Replace regularly. This will keep snails and other craweling insects from visiting your garden. Ocean sand works good too.

Diatomaceous Earth (Dia-earth - DE)

Diatoms (Diatomite) are either lacy-like snowflakes in appearance or tubes, which are better. They are microscopic in size. Their color is usually somewhere between white and brown. They are found both at the bottom of the ocean and in fresh water lakes.

Their by-product is oxygen. Three-quarters of the air we breathe comes from diatoms (they carry on photosynthesis). As the diatom dies, the dia-earth (dia-earth is 90 percent pure silica) falls to the ocean bottom or lake in which they lived. Layers of this silica form and do not degrade. After 150 millions years of this, we now have rich deposits of this substance.

DE is chemically identical to quartz and ocean white sand. DE is totally harmless to mammals if eaten and is recommended to be given to your animals as part of their diet. Many years ago when tests where run on the toxicity of DE, it was found to be beneficial to the animals fed with it as they gained weight and seemed healthier then those not fed DE. Animals fed DE have fewer problems with intestinal parasites.

DE can be fed to animal with their food, or in a bucket mixed with a little grain. For cats, use one tablespoon mixed along with their wet food, once per week. For dogs, add 2 tablespoonful per week mixed well with their wet food. For horses, use one-half cup per bucket of sweet oats, or add to a bucket of drinking water.

One more benefit of animals eating DE is that it removes worms, most intestinal parasites, and also effectively controls flys since the DE kills their grubs and reduces the flys population. DE contains many vitamins and minerals so horses love it!
Give weekly for best results.

AVOID BREATHING

Breathing DE should be avoided because it is rather harsh on your lungs,like any irritant. Try breathing flour is you will get the same results. So it is best to keep the dust down to a minimum. DE is not harmfull to your health if used properly. Becareful when using DE.

Keep it away from your eyes. De is like millions of tiny razor blades, and should be washed out with water immediately. Avoid rubbing your eyes, for obvious reasons. Wear a face mask when using for extra protection. You can use duster Mizer to apply(available at most mail order companies such as Nitron), or you can use a flour sifter or pest pistol.

DE prevents snails from laying their eggs and kills them by slow dehydration. It will significantly reduce the snail population over several years. Remember, the best results may take longer but are more permanent.

DE should be dusted onto plants and placed as a barrier around them. Avoid water for the next 24 hours to allow the DE to take effect.

AN IMPORTANT NOTE ABOUT DE

Many have asked me if the DE which is used for pools is the same DE I am always talking about. Well, it is and it isn't. It comes from the same geological sources but the pool companies use sulfuric acid to melt the DE crystals into a form which works best as a crystalline silica filter.

This 'melted " DE is more dangerous to humans if it is inhaled. Do Not use this type of DE in your garden. The garden variety is non-crystalline and can be dissolved by natural body fluids.Of course it is best to avoid breathing any type of dusts whether it is flour, DE, or whatever. Always thoroughly wash your face and your hands.

How Diatomaceous Earth works

Dia-earth(DE) is a dust that bugs attract by way of the static electricity produced when they move about. Nothing 'Bugs" an insect more then being dirty, hence cleaning themselves is a major part of their daily activity. DE clogs their breathing 'holes' (insects breathe through their outer exo skeleton), and removes oils that insects naturally produce to help protect themselves, leading to dehydration and death. This happens slowly, over a period of time that varies depending on which insects are affected, and on various environmental conditions.

Usually a small amount will do. It is for this reason that farmers use very little DE as they want faster results and are impatient. DE should be applied as part of your annual maintenance work. DE can be blended with water and part soap and sprayed on the plants or insects. When it dries, it leaves a thin coat of DE behind.

See Resource Directory for other Sources of DE

An Organic Snail Spray

10 drops SuperSeaweed per gallon
or 1 cup liquid seaweed per gallon.
1 cup Diatomaceous Earth per gallon.
1 cup Nitron A-35 per gallon.
1 tablespoonful Bio-degradable soap per gallon.
1 tablespoonful Tabasco Sauce or Cyclone Cider

To 1 cup DE, add 1-5 tablespoonsful tabasco sauce or Cyclone Cider, add 1-5 tablespoonful Biodegradable soap such as Shaklee's Basic H, Amway's LOC or Dr. Bronner's Peppermint soap. Slowly add enough water to form a slurry. Pour this mixture through a strainer into your gallon sprayer. Add 10 drops Superseaweed per gallon or you can use any natural seaweed concentrate available in your area or through mail order and 1 cup Nitron A-35 or another other natural bacterial product. Spray this mixture once per week(during infestation) in vegetable gardens, around lawns, fruit trees,etc. Spray plants and areas around. For best results spray in late afternoon.

See resource directory for more listings, see also chapter on making your own organic sprays.

Always test plants, crops, to be sprayed to see adverse results before spraying larger areas! If Snail infestation is heavy, double the DE and Nitron and spray more often. Be careful with spraying delicate flowers, and leaves of plants. It is best to spray under plants, on the lawns, under trees (on tree trunks). Do Not Water for 24 hrs to allow max effects.

Alfalfa Meal

Alfalfa meal is high in nitrogen and other natural bacteria and makes an excellent snail barrier while feeding your plants at the same time. Sprinkle around base of plant, do not water for 24 hrs.

Kelp

The use of Kelp or any dried seaweed in your plan to control snails is highly recommended. Kelp will provide the soil with a great many trace minerals; all essential in maintaining a balanced soil. Kelp will also reduce the snail population by raising the salt level too high for them, yet is tolerated by the plants and soil. Use various different types of seaweed. Different parts of the world produce seaweed that is rich in different trace minerals, so the various kinds of seaweed complement each other.

See chapter on Making your own organic sprays for more information on seaweed.

Superseaweed(or your own liquid seaweed blend) will be useful here as it is a blend from around the world. Spray the soil and plants with the Superseaweed once or twice a week to make certain that you have all the trace minerals that your soil and plants need.

IF you live near the ocean you can collect your own seaweed,but you must allow it to dry properly, separating the salt for later use. You should test the seaweed as well as the salt for toxins, because the ocean is getting more polluted. Also ask for test results on any seaweed product on the market today.

See Resources Directory for some excellent sources of seaweed

Mulching

Mulch is a highly recommended form of snail control. Its effectiveness depends upon the type of mulch you are using. I recommend alternating between pine needles, kelp, rock dust, various types of leaves, compost (makes a good mulch too if made with bark chips), greensand (too expensive as a mulch but I added it here because you should add a thin layer once per year), flour(yes, regular flour makes a great mulch/barrier!), bone-blood meal(a thin layer once per year will do it), and any other assortment of available mulches.

The more varied the mulch the greater the success. Look around in your area for free available mulch. The easiest available mulch is old horse manure which is at least 9 months old and has no smell at all.

This makes a great mulch to use provided you use assorted organic fertilizers as well as plenty of peat moss or pine needles to keep the PH down to around 6.5, depending on what you are growing. **Page 19**

NATURAL PEST CONTROL

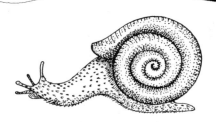

Edible European snail,

Making your own Sticky Stuff

1 cup petroleum jelly
10 oz castor oil
1 oz Cayenne Pepper
1/2 oz Tabasco Sauce

You can make your own Sticky Stuff by using petroleum jelly and castor oil, blend the two together and apply to trunk of plant. You can also add 1-2 0z cayenne pepper and or 1/2 oz of tabasco sauce to this mixture to increase effectiveness. You can make Tangelfoot work better by mixing in Cayenne pepper.

FLOUR or CLAY

Flour or clay can be used also as a barrier. Use 1 cup flour or clay add enough water to make a slurry, add 5 tablespoons cayenne pepper and stir well. Can be painted on trunk. or use as a dust around plants.

A Bio-Dynamic Snail Barrier

Old time bio-dynamic farmers use old Horse manure mixed with clay and seaweed painted as a barrier around trees. To do this add enough old horse manure to 1/2 fill a bucket. Add 1 or 2 lbs clay or flour, 1/4 lb cayenne pepper, 2 lbs powdered seaweed. Add enough water to make into a paste. Paint onto trunks.

COPPER

Copper bands wrapped around the trunk of a tree or plant also comprise a good, non-toxic form of snail and slug control. Snail Barr, a product designed for this function is available from Custom Copper(see resources). Copper clips and barriers can be used in your garden by fastening them to raised beds, and branches.

Natural Predators

There are many natural snail predators. Birds love to eat snails which are a very important part of their diet. Beetles prey on snails,slugs and other insects. Toads are especially fond of snails and slugs.

A Pond can be a good tool in attracting toads to your area. Many snakes and lizards love snails and slugs.

Decollate snails are a very effective natural predator against the brown garden snail. This snail is smaller and has a shell that looks like a seashell. This snail prefers decayed vegetation and small snails are its favorite diet. Larger snails will have to be hand picked. When using decollate snails be careful in using any other methods described above as they will effect these snails also. Use only as a last resort.

Chickens love snails....hint
Snail Predators are available from:
ARBICO (Arizona Biological Control, Inc.) ,
Peacefull Valley Farm supply
See Resource Directory for address

IPM Snail Control Dust

1 lb Greensand or Rock Dust
1 lb Dia-earth,
1 lb Kelp meal
1 lb Alfalfa Meal

Blend together. use a dust or around base of plants. See Chapter on Rock Dust for more information. While DE is good for animals, it is hazardous to snails. When handling, avoid breathing it in and getting into eyes as it is rather harsh and if this occurs, wash out your eyes immediately with copious amounts of water,

Remember that it will take time to control the Snails.

Stop using chemicals that kill of benificials in the soil.

Use more compost and mulch to regain a eco- balance of your soil.

Invite other creatures into your garden.

Chapter 3

How to Control Gophers Organically

Pocket Gophers, or ground squirrels as they are sometimes called, can be as big a problem as ants. Gophers are like little bulldozers. They dig a network of tunnels usually six to 12 inches below the surface. Tunnels near the surface are for gathering food; deeper ones are for sleeping, raising the young and storing food. Well-suited to burrowing, gophers have small eyes and ears that don't clog with dirt, and sensitive whiskers and tails to guide them in the dark.

Gophers have long teeth for biting off chunks of hard soil or roots. They are sensitive to sound, and a variety of smells, as well as light. To successfully control gophers and other rodents, one needs to understand their senses of smell and hearing.

If you see a plant being dragged under ground you know you have gophers. Gophers eat plant roots as well as bulbs. They will eat almost anything including most vegetables such as tomatoes, carrots, squash, lettuces, etc and they also love roses. They'll nibble on your fruit trees, most flowers, and tulips.

Moles are insectivores they eat earthworms, bugs, larvae, and only occasionally nibble on greens and roots. They do very little damage other then through their tunneling, unless there is very little food or water.

Gophers can do a great deal of damage but are an important part of the natural ecosystem. I do not recommend the use of traps as they are inhumane and very cruel.

It is important that you properly feed and care for your plants as gophers are less likely to attack healthy plants. By top-dressing your lawn and property with year-old horse manure, you will not only be feeding your soil, but also repelling gophers naturally as they do not like manure. You will also be encouraging natural predators as compost promotes healthy environments.

The Process

Day One

When you have spotted the telltale signs of gophers a small mound of soil or dead plants locate its entrance and remove enough soil to widen the opening. Then put a hose in and turn on the water. Allow to run for at least 10 to 15 minutes. Look around for any water or gophers coming out. The water coming out tells you where the tunnel leads to.

During this time you might drown the gophers. But this is not the primary intention. The purpose of the water is to encourage the gophers to move to another site. You can force them off your property in this way.

Remove the hose and put a cotton ball soaked in citronella oil, garlic oil and onion extract in to the hole. Wear gloves to disguise your 'human' scent. Cover up the entrance and pat down the soil. Repeat at all the locations where gophers 'mounds' are present. Return the next day and find the tunnel where the cotton swab has been thrown out of its tunnel. That is the active tunnel.

Second day

Place a mixture of the formula listed below in the active tunnel, then cover the entrance. Cayenne Pepper is a natural irritant and gophers will not like this. Garlic pepper will work well too, as will tobacco dust (see formula).

A good device to inject the mixture into the gopher tunnels is the Wilco Gopher Getter, Jr., like one which is available from Peaceful Valley Farm Supply. It is lightweight, and easy to use. The baiter applies bait directly into the tunnel as it is pushed into the gopher runway. It places bait quickly and safely where gophers are. It comes with 685 grams of poison bait(tell them that you only want the empty container and not the poison).

This is a bait made from milo grain and strychnine, a natural highly toxic poison derived from Nox Vomica. a nightshade.

This is the most poisonous material Peaceful Valley has so EXTREME CAUTION IS URGED. It is natural, but also highly poisonous against gophers and moles. Keep the warnings in mind so as not to unintentiionally poison any birds, rodents, animals and people. Must be placed in gopher run, away from where birds and others can get at it. One grain can kill a bird!!

READ THE DIRECTIONS AND HANDLE AND STORE WITH CARE. Contains .35% Strychnine Alkaloid. I do not recommend that this bait be used. Instead I would ask the folks where you buy it to remove the poison and to sell you only the empty container Or when received that you carefully throw away the poison and wash out the container. Wear plastic gloves! Please be very careful with this stuff!.

NATURAL PEST CONTROL

Dispose of properly

Inside your container mix enough cayenne pepper, or tobacco dust as in formula, to fill the container which is screwed onto the bait applicator. This will be enough to give them a very bad headache and will keep them away. Again I do not recommend using any poison especially if you have children,dogs or cats. Wash hands after use and wear gloves while you handle this stuff. Be very careful.

Day Three

This is the day you create the infamous Gopher Sneezer. Thoroughly mix together equal amounts of tobacco dust, cayenne pepper and dia-earth. Place this concoction into the gopher tunnels with the injector. For any new gopher tunnels repeat the process of placing the cotton ball soaked in garlic, citronella oil and onion extract and cover the tunnel entrances. Once you know that the particular hole is active, repeat the process of applying the gopher mixture into this new tunnel. Inportant: Do not Water for 24 hours. Repeat as often as needed.

Vitamin D?

A form of Vitamin D, kills gophers, rats, ground squirrels, mice. Rodents have trouble regulating calcium in their blood stream, vitamin D disrupts their systems, ending in death. Additionally, using vitamin D is safe for household pets and humans. Add to your injector and place into gopher tunnels only after other attempts have failed. Available in pellets or granules. Available from Gardens Alive.

Barriers

The use of barriers is recommended in controling gophers around your home. Use fencing that is woven tightly enough to keep the gophers from going through and strong enough to resist their chewing. Place the fence underground, around an area where you plan to plant a garden.

Also, use plants that gophers naturally abhor. Garlic and onion plants are avoided by gophers and make excellent barriers. Other plants gophers hate are herbs such as tansy, peppermint, spearmint and rosemary.

Gassing out the Gophers

Gassers that are often used to evict gophers from your property are made of sulfur and potassium nitrite. They are lit and placed in the tunnel, producing large amounts of smoke which drive the gophers away. They must be used regularly to be effective. Avoid using this method unless you know how to use the gasser.

Great care must be exercised because the sulfur smoke is very dangerous to inhale and will cause serious illness. Available at most nurseries. Avoid using if you have kids.

Hair Today Gone Tommorrow

Another 'gasser' is decomposing hair. A strong odor is given off as the hair decomposes. Placing human hair in the gopher tunnels effectively keep them out of the passages and should be placed in new tunnels as they appear. Hair works best on hot days and nights: The heat causes the hair to smell up the tunnels.

GOPHER ROCKS

Straight from Europe, a product called Rodent Rocks are porous lava stones that have been soaked in an herbal formula containing garlic and onion. While not necessarily considered gassers they do smell up the tunnels. When buried, the rocks give off a very strong odor that effectively repels most rodents (including gophers, moles, mice). Bury the small rocks (six inches deep and two to four feet apart) in a protective circle around your vegetable garden or problem area. Rodent Rocks remain effective for four to 12 month. This product is widely used in Germany. I recommend this in place of poison bait if you have animals.

Using fish heads in gopher tunnels will irritate them into moving on! Get the picture? use any thing natural that has a foul odor. Cotton balls soaked in ammonia also works well.

GOPHER IT!

When placed in the ground, the electronic stake vibrates and emits a noise in 15 second intervals, causing underground dwellers within a 100 ft. diameter to flee the area. The manufacturer claims that it is effective against gophers, moles and ground squirrels, as well as shrews, voles, pocket mice and kangaroo rats. Gopher It eliminates the need for traps, poisons or gases. Totally safe for use around children and pets. Actual size is 12" long. It comes with on on/off switch and requires 4 "D" cell batteries (not inlucded).

NATURAL PEST CONTROL

Animals

The use of animals is the best way to naturally control gophers. If you have animals, the use of poisons is not recommended. Cats and Dogs will chase gophers, and sometimes eat them.

Other predators that will eat your gophers for dinner are king snakes, gopher snakes, hawks, coyotes and owls. Try buying a gopher snake or King snake from your local pet store and releasing it down the gopher hole! A king snake is also a natural enemy of the rattlesnake. Try building a nesting home for owls and other birds.

TRAPS

Gopher Purge Plant

Uphorbia Lathyrus, Each pod contains 3 seeds. Repels gophers and moles and other tunneling and burrowing animals. The roots are extremely caustic to them as well as to humans so be careful not to get the white milky stuff on your hands or face, wash off right away!.

Gopher Purge is a parental and gives plenty of seeds for replanting. Water well at start. The whole plant can be dried and the herb placed into the gopher tunnels as well as many other uses for it.

For maximum protection it is important to plant Gopher Purge as a hedge that surrounds the property if possible. Planting Gopher Purge(a complete seed pod should be planted) every 6 feet the first year and then decreasing it to every 3 feet the second year then down to every 2 feet if the gophers are still bad. This is the safest method of ridding yourself of these gophers. Give it time; it works!

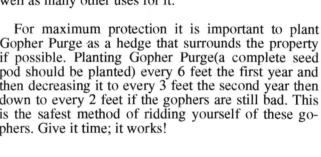

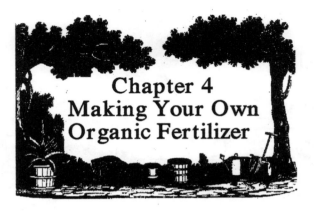

Chapter 4
Making Your Own
Organic Fertilizer

Growing pest free vegetables and fruits is in part dependent on not using chemical fertilizers and instead to rely upon the various natural resources available to us. It is not within the scope of this book to debate the difference(s) between organics and inorganics; nor is not my intent here to show why you should not use chemical fertilizers or if a plant can tell the difference bewteen organic nitrogen sources and chemical sources. I am not here to convince you that you should go organic and not chemical.

These things you must have come to the conclusion on your own. Instead I am here to pass along to you my knowledge of Organics. In this chapter, I will cover some basic ways that you can make your own organic fertilizers. Please note that there are more and more organic fertilizers available on the market today, not all are 100% organic. Read the labels very carefully.

The physical structure of what we feed the Earth cn be modified. For example, a rock can be crushed into powder, or garlic made into a liquid. All these things are still subject to the laws of nature. They are biodegradable and non-toxic.

This information is but a small example of the various resources we have to make our own fertilizers. Time and experience will teach us what works and what does not. There are many books available on composting, such as Backyard Composting by John Roulac, and those available through Rodales books (see resources directory).

"High Nitrogen fertilizers contribute to many insect and disease attacks."

Natural and Organic Elixirs

Here are some formulas for organic fertilizers. I have included them to help you develop a feel for blending your own organic materials. Remember that the numbers only signify what is available immediately, not what will become available later on as nutrients are constantly released into the soil by the action of the micro-organisms in the soil. Also, you may adapt these formulas to suit your needs and experience.

For Fruit Trees

The following recipe makes approximately 30 pounds.

2 lbs fish meal
2 lbs Alfalfa meal
4 lbs cottonseed meal (use only organically grown)
1 lb wood ashes or trace minerals
2 lbs rock dust
2 lbs seaweed meal or kelp meal
2 lbs coffee grounds
10 lbs Compost
5 lbs Composted animal manure(llama, rabbit or best)

Blend all ingredients well together. Make only enough for immediate use. Keep dry. Spread a thin layer under each tree. Start from one foot out from trunk to twenty feet past the drip line. Water well into soil. Use approximate one-half pound per year of growth. A 10-year-old citrus tree should get five pounds twice per year: once in the spring and once in the fall. Apply lots of compost or horse manure in early spring (always add soil acidifier such as pine needles, leaves or compost to horse manure in order to lower the pH level. Most fruit trees prefer 6.5 pH). Never use freshly cut grass clippings as a mulch.

Spray leaves with a liquid seaweed solution (like Superseaweed) or with a mixture (see chapter on making your own foliar spray) once per month. Adding small amounts of soap will control pests. See soap page.

For Roses

The following recipe makes 30 pounds. To make less follow proper proportions.

2 lbs fish meal
4 lbs Alfalfa meal
1 lb cottonseed meal(if soil is alkaline)
1 lb wood ashes (if soil is very acid)
2 lbs rock dust
4 lbs Kelp meal or seaweed meal

a dash of garlic powder
1/4 lb organic tobacco dust or smoking tobacco(optional)
10 lbs Compost
5 lbs composted animal manure

Blend all ingredients well together. Wear rubber gloves when handling the tobacco dust. Smoking tobacco (available from Santa Fe Natural Tobacco Co. in Santa Fe New Mexico, see resources section) is the best tobacco to use for this purpose as it is organically grown.

A blend of garlic and tobacco will keep many pests from attacking your roses (try growing garlic at the base of roses for even better results). Fertilize once per month during main flowering season. Use one cup per plant, per month. Water in well without wetting leaves. Use only filtered water (see resources for garden filters).

The compost should have an acidifier added to it, such as pine needles, etc., to bring the pH down to six. Roses grow best at a pH level of 6.2. Roses love lots of rich compost and mulch. In early spring, add enough compost to fill well around roses. Water in well before mulching with leaves, bark etc. Roses grow best with a drip system. For best results, spray weekly or monthly as needed with a liquid seaweed and soap mixture. See chapter on making your own pest controls for additional rose sprays.

For Azaleas

The following recipe makes about 25 pounds.

4 lbs Cottonseed meal (organic)
2 lbs Rock Dust
4 lbs Alfalfa Meal
1 lb Kelp Meal or Seaweed powder
4 lbs coffee grinds (fresh coffee can be used)
10 lbs compost (with equal amounts of aged wood)

Blend all ingredients together well and use as needed. Azaleas love acid and require a low pH level of 5.5 to 6.0 for best results. Always add an acid aged wood to your compost when feeding azaleas.

For the Lawn makes apprx 45lbs.

10 lbs Fish Meal
10 lbs Alfalfa Meal
5 lbs composted animal manure (chicken, steer,rabbit or llama are good here due to high nitrogen content.)
10 lbs rock dust
1 lb wood ashes (if available)
10 lbs compost

Blend all ingredients together well, and apply four times per year, or as needed. Sprinkle thin layer over lawn and water well. Lawns love compost, so top dressing lawns once per year with a good rich compost will keep the soil alive and allow for deeper root systems and a pest-free lawn. Compost should be raked into lawn and then watered well. The best time to apply compost to the lawn is in early spring (or as soon as last frost is over). For best results, spray your lawn monthly with a liquid seaweed.

For African Violets makes apprx 8 lbs.

1 lb cottonseed meal
1/2 lb rock dust
5 lbs compost
1/2 lb Alfalfa meal
1 lb kelp meal or seaweed powder

Blend all ingredients together well and apply using a tablespoonful per small container plants or add a tablespoon of the mixture to a cup of water, sprinkling it on the violets.

For Container Plants makes apprx 13 lbs.

10 lbs Compost
1 lb rock dust
1 lb Alfalfa meal
1 lb kelp meal or seaweed powder

Blend ingredients together well and apply mixture as needed to container plants. Usually one cup per 5 gallon container will do, but the exact amount will depend on the size and the age of the plant.

For the Vegetable Garden, makes apprx 27 lbs.

10 lbs compost
5 lbs rock dust
2 lbs kelp meal or seaweed meal
5 lbs composted animal manure
5 lbs Alfalfa meal

Blend all ingredients together well. Feed plants weekly or monthly as needed during growing season. Vegetable gardens love good rich compost. Apply compost as both a top dressing and mulch.

Page 25

Making Your Own Organic Fertilizers

Feed flowering plants the same as Roses.

For Sick Plants

10 lbs compost
2 lbs rock dust
2 lbs Alfalfa meal
2 lbs kelp meal or seaweed powder

Avoid high nitrogen fertilizers. Spray with a liquid seaweed concentrate such as Superseaweed. Apply it as often as needed.

Don't Want to Make Your Own Compost?

Look around your city. See if you can spot any places that sell compost. Ask other gardeners for a source. Go out into the country side and look around there. Allways ask them what they are doing with it and how they are making it. Ask them if they are exposing it to any chemicals etc. Never assume that anything is as they say it is. Ask about how they make it and what goes int it, how long have they been making it and so on. These are important questions which you should be asking. Now a days there are more and more people making and selling compost so it should not be hard to find.

"Avoid buying compost made with sewer sludge since it may contain heavy metals which pollute ourselves and our environment."

Here are some suggestions of possible sources of compost materials or compost all ready made

Talk with your local restaurant stores especially the natural food ones as their vegetables, etc are organic.

Look seafood locations where they either cook them as in a restaurant or where they catch they as in a dock. These can be dried for your compost use.

I also suggest that you collect clam shells etc as these when dried and chrushed make an excellent addition to a compost pile.

Look for any sources of coffee such as either coffee hulls and or coffee grinds,similarly, locate sources of tea, either tea bags or leaf.

Lumber yards may also be a good source of material provided that you only use untreated wood products.

Locate chicken farms, ask for their feathers and or shell as well as any droppings. Ask them what they are spraying if any thing.

Locate horse or cattle farms ask if they spray their animals or the manure.

Look for an organic mushroom farm. make sure they are organic(use steam instead of chemicals).

Look for old alfalfa bales.

Grass clippings make an excellent source of nitrogen for your compost pile, just make sure its organic(unsprayed).

For more information see compost chapter.

Various Types of Organic Fertilizers

Natural fertilizer	N	P	K	Comments
Alfalfa Meal	3	1	1	minerals and nitrogen.
Bat Guano	3	8	1	High in calcium, minerals
Blood Meal	12	2	1	Hign in nitrogen
Bone Meal	3	12	0	25% calcium
Earthworm Castings	5	2	2	many minerals
Feather Meal	12	1	1	good for lawns
Fish Emulsion	5	1	2	avoid Urea based
Fish Meal	10	2	2	a high nitrogen source
Greensand	0	2	5	trace minerals slow realease
Kelp Meal	2	1	8	best for all use
Rock Dust	0	1	4	high in calcium, iron, etc
Seaweed	2	1	5	great as a foliar feeder
Soft Rock Phosphate	0	16	0	high in calcium
Sul-Po-Mag	0	0	22	22% sulphur, 11% magnesium

Chapter 5

Making Your Own
Organic Pest Controls

The increasing use of synthetic chemicals in our daily lives is causing an increasing imbalance in Nature. When nature is out of balance it causes stress which in turn causes disorder and chaos. It is during this period that diseases and pests will strike. The plant and insect kingdoms are but mirrors into our world. More like a reflection, we seen the results of our actions within the insect and plant worlds. How long can we ignore the chaos?

It is for these reasons that natural non-chemical lifestyles are are very important in developing a sustainable lifestyle. This has created a demand for the knowledge of how it is done.

This book attempts to pass along some of this vast information. Use it, experiment with it, discover and let others know. Sharing your knowledge with the world insures survival.

There are many books on the market that will help you control and eliminate chemicals from your life. Debra Lynn Dadd has many books and a newsletter towards this end. See Resources in the appendices for address. The Invisible Gardeners of America is a club I started for this purpose concentrating on natural pest controls, compost production, food production, natural pet care, organic horticultural techniques from past to present are all covered. See Resources for address.

Fortunately for the organic grower, there has been an endless amount of knowledge on organic controls passed down through history. The science of organics is the understanding of the delicate balancing act which Mother Nature does every day. It is understanding this balance that makes organics work. Organics has been used in pest control and farming since before written times. Chemicals are synthetic, temporary and unstable. Organics are natural, permanent and stable.

But although non chemical sprays and dusts continue to play a crucial role, biological controls are of equal importance. By attaining a thorough understanding of the relationships between plants, soil, animals, insects and man, we can begin to understand the deeper relationship man has with nature.

The Miners Canary

This is a story of how the miners would take a bird in a cage down with them into the mines. If the bird died then everyone would leave since something was wrong. The same is currently happening to us.

The disadvantages of chemicals are well documented with the result being the depletion of top soil (what's left is nothing but sterile and lifeless) and the extinction of many species of life; to the mutation of pests into greater pests, and the weakening of all living beings (environmental toxins cause eventual death and extinction). The depletion of the ozone layer is but one example of this.

NATURAL PEST CONTROL

"This is the way the ancient farmers did it, by relying on natures methods of maintaining balance."

Some Fundamental Rules of Organic Pest Control

Rule #1
The Higher the Energy,
The Greater the Balance;
The Lower the Stress,
The Less the Pest.

High energy soil provide greater balance which reduces stress and reduces the pest activity. All Pest activity is linked to high stress levels. The greater the stress the less nutrition is available. Less nutrition increases stress! Another way to put it is: The Greater the stress, the greater the pest.

Rule #2
Everything is Linked Together.

What happens to one affects the other.

There is a Chain that links us all.

"The Frog does not Drink up the pond in which he lives" A Buddhist Proverb

Rule #3
Healthy Soil Equals Healthy Plants.

Providing a bacterially and minerally balanced soil insures correct nutritional availability to plants. Well nourished plants develop stronger immunities to insect and disease attacks.
Pests attack plants that are stressed out..

Rule #4
Strive to achieve Ecological Balance in your Environment.

Avoid using any thing that causes imbalance. All Chemicals cause imbalance, upset the environment and increase stress.

Rule #5
Use only Organic/Natural Fertilizers and Pest Control.

The Rule of Stress applies to all living things. Chemical fertilizers and chemical pesticides cause imbalance. Avoid high nitrogen fertilizers. Don't use Urea based products! Urea kills the beneficials, and stays in the soil for years. To have healthy soil you must have a diversity of bacteria and minerals, which is not possible when using chemicals.

Chemicals kill earthworms, destroy bacteria, lock up minerals in the soil, and cause illness through weakening the biological systems, opening them to attack from diseases or pests. High nitrogen causes rapid plant growth and tender plants.

Rule #6
Insects Develop Immunities to Pesticides.

Rule #7
Whenever possible, Avoid Killing.

The idea that pest control means killing the pest has developed through the high pressure tactics of chemical sales people. To sell more and more chemicals is indeed their goal. Killing upsets the natural balance. Let Mother Nature do this for you. She has already set up a system of her own that works. There are many ways to control the behavior of pests. Only as a last resort do you kill pests.

Rule # 8
Follow The Law of the Little Bit
Use less whenever possible.

Making the Organic Transition

Proper nutrition is the very foundation of organic growing and pest control. Just as a human body that isn't properly fed is wide open to attack from one illness or another, so it is with plants. Chemical fertilizers are not complete foods. They lack almost all of the important trace minerals. There are over 70 trace minerals that plants and humans need, and chemical fertilizers provide only a few of them. If plants do not get the trace minerals they need, as well as a variety of enzymes and bacteria, they will become stressed and sooner or later will be attacked by disease, or pests.

Ants, like most predatory insects, attack plants that are sick or stressed. This is a natural law, like the wolf who only attacks sick sheep.

I do not recommend the use of chemical fertilizers. If you presently are using these fertilizers and want to stop, be forewarned that you must proceed slowly and carefully as you make the transition from chemical fertilizers to purely organics. As an addict depends on drugs, plants depend on chemical fertilizers for their 'hits' and won't accept organic food at first.

Time is needed to allow the plants natural systems to begin to work again.

Give yourself and your plants at least a year to go from chemicals to organics. It may take several years of growing organically for the soil to return to life and for the organic materials to kick in. Use the following schedule in making the organic transition.

"It's not the Plants that are Hooked on the Chemicals"

First Month

Keep using the same amount of chemical fertilizer but also spray a liquid seaweed blend (such as Superseaweed, or Nitron) as a foliar feeder. Make sure the soil is in good condition by applying lots of rich alive compost. Stop using any chemical pesticides. Start using the many methods mentioned in this book.

Second Month

Reduce the amount of chemical fertilizer by one-third, while continuing to spray the leaves of your plants with the liquid seaweed according to the instructions (Superseaweed is five drops per gallon). Replace the chemical fertilizer with a good compost such as Organa. Apply more compost around plants, turn over soil when possible.
Mulch well.

Third Month

By now you should have reduced the amount of chemical fertilizer to one-half the original amount you were using before and have increased the amount of organic fertilizer to replace the chemical fertilizer. In other words, you are now using one part organic and one part chemical.

You will find that during these months your plants will be under a great deal of stress and may be attacked by various predatory bugs. Use only organic methods to get rid of them. Keep spraying the leaves with a good liquid seaweed weekly or monthly.

NATURAL PEST CONTROL

Fourth to Eighth month

Stay on the 50/50 basis. This is a good time to see how the plants are responding to their new organic regime. Some plants will not like it at all and may even die, but the majority of plants will do very well. Keep spraying the leaves as often as recommended (as above).

Ninth month

Reduce the use of chemical fertilizers by one-half again, and increase your organic fertilizer/compost. Follow with a monthly spraying of liquid seaweed.

Tenth month- twelfth month

Repeat the process. By now you should be using one-quarter of your chemical fertilizer with three-quarters organic fertilizer. This is the level you want to stay at for the next two months. By the end of the first year you should have reduced your chemical usage by 80 percent. You should also have increased your organic usage by 80 percent.

During the next few years you should work on reducing the chemicals used by 1/2 then eventually not using any chemicals at all and be 100% Organic!

Organic Controls

Any problems your plants may have can be treated with the appropriate organic controls.

Liquid Seaweed

Liquid Seaweed is very important because it provides plants with the necessary trace minerals to build up their immune systems. It also provides bacteria for the plants' natural systems to work with. Use distilled or spring water or well water for this as city water may kill off the bacteria (or you can filter the water. See garden filter appendix.).

Read the chapter on organic fertilizers for more details on what organic fertilizers to use. Remember that the first line of defense is careful planning. Many plant troubles blamed on diseases and pests are really caused by poor soil. By providing plants with healthy soil to grow in, you will find that plants will be healthier, and less prone to attack from diseases and pests.

A good organic grower plans ahead. He or she works into his soil as many different varieties of organic materials as he can get. He or she adds manure and mulches and provides the soil with as much and varied sources of nutrients as possible. This is the key to successful organic growing as well as to organic pest controls.

Proper Watering Methods

If a plant is improperly watered, it will suffer stress and open itself up to attack by insects and rodents. The lack of regular watering causes nutrients to disappear and become unavailable to plants. Learn the correct methods of watering. Too much water is just as bad as not enough.

Insufficient lighting also contributes to stress. Trees weakened by improper watering, root damage and improper feeding are more likely to be attacked by pests such as borers than are healthy trees. Bugs and diseases act as censors in nature. Bugs like unbalanced soil.

"Handpicking is one of the best Organic Pest Control methods. This means that you must first know your pest"

Get a hand lens and examine the stems and leaves and fruit surfaces to see what is causing the damage. Look for tiny eggs, tiny puncture holes, trails, excrement and insects. In other words identify the problem.

There are many good books on insect identification on the market. A good one is Natural Insect and Disease Control by Roger Yepsen Jr (Rodale Press, 1984). Another good book is Common-Sense Pest Control by William and Helga Olkowski and Sheila Daar (Taunton Press, 1991). Not all insects are bad, so be careful which ones you place the blame on.

Also learn to identify insect eating habits. Some insects are chewers such as caterpillars, earwigs and leaf miners. Others are suckers such as aphids, and mites. Still others feed below ground such as nematodes, maggots and rootworms, and others are such as apple maggots, beetles and tomato fruit worms.

Know your Insects

Each insect requires a different method of treatment. Diseases also demand various different methods of treatment. Learning how to recognize different types of insect damage and their associated diseases takes time and experience.

There are many books available which cover this subject of identification. Rodales Organic Gardening Association has many fine books on this subject.

Physical barriers are an important tool for the organic grower.

The Plant Guardian (TM)
IPM Formula

10 parts Dia-earth
1 part pyrethrum powder
1 part organic smoking tobacco or tobacco dust (optional)
2 parts Kelp meal
1 part African Cayenne Pepper
1 part rock dust or greensand
1 part Alfalfa Meal
1/2 part Sulfur or Copper dust
1/2 part garlic powder.

Mix all ingredients together well. Depending on how much mixture you need, use a cup for the part. Wear face mask and mix slowly. Use when needed around plants. Can be used with Pest Pistol or the Duster Mizer (see resources for address). This is a special formula developed by The Invisible Gardener. You must make your own as this stuff is not for sale. Potent! Use as a dust around plants.

This forumla can also be used to create a paste (slowly add water to mixture, and stir until it turns into a paste). Paint the paste onto plants with a paint brush. It will last a long time and will protect plants from crawling insects. Re-touch with paint brush when needed. Great for painting raised garden beds to protect flowers and vegetables. The paste will kill many insects that come into contact with it. **The paste also will kill beneficial insects** as well, so use on spot applications only. It is best for crawling pests such as snails and slugs, spiders, snakes, rats, gophers (place

into tunnels), ants, fleas (outdoors only), and will keep, rabbits, deer and coyotes away (sprinkle around property).

Tangelfoot

Is made from castor oil and bees wax. Very sticky stuff. Place around trees as a barrier. (I like to mix the Tangelfoot with Plant Guardian to make an even more potent barrier).

Tabasco Sauce

Makes an excellent barrier to many insects. Simply add one tablespoon of tabasco sauce into a quart sprayer. Spray on an area you want insects to stay out of. When it dries it leaves a residue behind. Another favorite of mine is to use clay and mix it with cayenne pepper. I use it to paint the trees or dust around the outside of the garden. Snails hate this stuff!

Traps

Japanese Beetle Traps are a pheromone(a natural insect attractant) that attracts females and traps them for disposal. Fly trap, see fly chapter. Apple maggot Traps,apple sized red spheres coated with a sticky substance. Attracts females maggot flies. Use Tangle-trap a brush on sticky substance for above trap.

Sticky Traps, comes in either yellow or white. Attracts and traps them.

There are many Light Traps, pheromone traps and sticky traps available for many different types of insects. A good source of these traps is Peaceful Valley Farm Supply, Harmony Farm Supplies, Gardens Alive, Necessary Trading Co, Gardener's Supply, Bio-Integral Resource Center or ARBICO, etc.

Biological Controls

The use of beneficial organisms offers an effective system of integrated pest management. Not all insects are bad. The good guys and the bad guys do not live in the same place. When the good guys move in, the bad guys move out. It's that simple.

Some insects help to improve the soil through aeration, while others help by generating compost from leaves and other decaying organic matter. Some insects, such as bees, also help in pollination.

Without the help of insects, we would be covered with our own waste. Insects recycle waste.

Biological Controls through the use of a friendly insect to get rid of the unfriendlies should be used in conjunction with an integrated system of pest controls:

NATURAL PEST CONTROL

monitoring, identification, trapping, planting pest resistant varieties, proper fertilization, proper irrigation and the use of natural insecticides.

Care must be taken in releasing biological controls to provide them with a beneficial environment. Provide for an environment that promotes a diversity of predators and parasites. A wide range of plants also is important. Before releasing the beneficial bacteria, water the area well. This provides for immediate sources of water for them. Release beneficials in evenings only since hot days would make it difficult for them to establish themselves. They need to find a safe hiding place.

Bio-Insecticides

When using BT, add a little molasses to increase effectiveness. A natural UV inhibitor will make it last longer and not break down from the sunlight. A good natural sticking agent will help here. Oils make good UV inhibitors.

BT

BT(Bacterium Thuringiniensis) discovered in 1911, is a type of natural bacteria found in the stomach of certain caterpillars . There are over 35 Bt varierties on the market. Use only when necessary as pests will develop an immunity to this.

BT, controls caterpillars, corn borers and mosquitos. A stomach poison that stops insects from feeding and become paralyzed. Effects only targeted insects not effecting the environment.

BT, Kurstake Bait, controls European corn borers. Apply early spring.

BT/San Diego M-One, controls Colorado potato beetles.

BT/H-14 Gnatrol controls fungus gnats. Fungus gnats lay eggs in soil. They love potted soil, peat moss and organic rich humus. The larvae feed on plant roots and root hairs crippling(if not killing) the plants. Population is highest during winter month and spring.

Infested plants show discoloring, wilted leaves, root rot. They attack ornamentals, many vegetable crops, poinsettia and others.

BT/Israelensis controls mosquito and black fly larvae. Adult females drop eggs in water, to hatch under water. 98% die after only 5 mins of low exposure to this product! Kills only larvae. Harmless to fish, birds, mammals, plants and the environment.

MVP

MVP(a natural bio-insecticide) is a new version of BT. Stays effective for as long as eight days. Effective against most worms such as armyworms, imported cabbageworm, cabbage webworm, cabbage looper, hornworms, corn borer, leafrollers, moths and most caterpillars. Will not harm mammals and beneficials, fish or birds. Use on cabbage, broccoli, root crops, lettuce, corn, tomatoes, peppers, citrus fruits, peanuts, apples and many more plants. Can be applied up to day of harvest. Use 1-1/2 to 3 oz per gallon. Available from Gardens Alive!, Peaceful Valley and AR-BICO.

Semaspore, Nosema locustae, for control of grasshoppers. A naturally occurring protozoan is deadly to grasshoppers and crickets that eat it but is pest specific and will not harm any thing else. Infected grasshoppers spread the disease to newly hatched grasshoppers. Apply in early spring for best control.

Beneficial Predators

Aphid-Lion

Chrysoperia Camea and Chrysoperia Rufilabris: Some aphid-lions are also known as dobsonflies, ant-lions, or lacewings. The aphid-lion is found in most gardens and are predaceous. They are an all-purpose garden predator. The larvae is eats aphids, mealybugs, scales, thrips, mites, spider mites and whiteflies. They destroy many other destructive insects, as well the eggs of many caterpillars, mites, scales, aphids and mealybugs.

Lady Beetle or Lady Bug

Many varieties are native to the United States. Both the young and adult stages eat various soft bodied insects such as aphids. If a good food source is available, the lady beetles will stay and lay eggs.

Dragonflies and Damselflies

The **Mosquito Hawk** is a good name for the dragonfly. They have highly developed eyes and a speedy mode of flight. They also are fierce hunters. The damselfy is the smaller of the two and unlike the dragonfly, folds her wings on her back. The young of both the dragonfly and the damselfly are called nymphs or naiads and they devour mosquitoes and other water-born insects.

Fly Parasites

Attack flies before they hatch with this parasite. Fly parasites deposit their eggs inside immature fly pupae, the parasitic eggs hatch

NATURAL PEST CONTROL

into larvae which feed on their hosts.

Praying Mantis

The Chinese mantis was first introduced into the United States in 1896. The mantis captures, holds and devours many different bothersome insects. A very helpful insect in the vegetable garden or around the house.

Spined Soldier Bug

Controls mexican bean beetle: This bug preys on many garden pests such as cabbage loopers,cabbageworms and mexican beetles. Both adults and nymphs attack other pests.

Semaspore

A long term grasshopper control which is deadly to grasshoppers and crickets, is made from a naturally occuring protozoan.

Predators

The most abundant and important predators for a healthy garden are the two-winged Flies, the Wasps and the Green Lacewings. Parasites will attack insects in all stages of their development. The host does not die immediately but provides nourishment until the larva of the parasite is nearly grown after which the host dies.

Tachinid Flies

These predators prey on a wide variety of insect species. They lay their eggs in the host body, which provides the young with a source of food. Compsilura concinnata is a fly that was imported from Europe to combat the gypsy moth. Flesh flies have varied habits: Some are flesh eaters, while others eat only insects.

Wasps

Wasps feed mostly on other insects. Their favorite is caterpillars such as the armyworm. There are many varieties of parasitic wasps. Wasps are generally quite social, with males, females and sterile workers making up the family. Some species of parasitic wasps attack only certain insects. Encarsia formosa is the whitefly parasitoid which effectively controls whiteflies.

Lysiphlebus Testaceipes

These bugs destroys millions of aphids per year by laying eggs within the bodies of the aphids.

Trichogramma Minutum and T Pretiosum

These are a minute egg parasite that destroy the eggs of most injurious pests such as the bollworm, cotton leafworm and various borers, hornworm, colding moth, and all moths and butterfly eggs.

Honey Bees

Honey Bees are very valuable in their role as pollinators. as well as a source of food.

Beneficial Nematodes

These parasites control many borers, grubs, cutworms, oriental beetles, pillbugs and cutworms. Available are parasitic nematodes of the HB and SC varieties. The Hb variety is an effective alternative to milky spore for the japanese beetle grub.

Clandosan

These parasites control root-knot nematodes, and simulate naturally occurring soil organisms. They should be tilled into the soil before planting, and then applied every year after that. They also can be used as a side dressing as needed.

Fire Ant Bait

Contains avermectin which is a natural soil organism. Ants take bait back to colony where it is distributed through out colony and kills entire colony within 3 months. Use the same system as on ants.

Here are some good sources of biological controls, predators and parasites: ARBICO, Bio-logic, Bountiful Gardens, Peaceful Valley Farm Supply, Rincon-Vitova Insectories, Inc., Bio-Integral Resource Center, Gardens Alive!, Mellinger's, and the Natural Gardening Research Center. See Resource Directory for addresses and more listings.

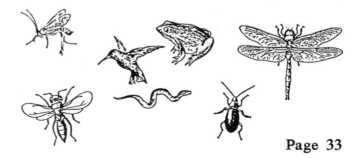

NATURAL PEST CONTROL

For the Birds

Birds eat many different types of insects year round. Providing housing, food and water for our bird allies will help insure good insect balance.

A few examples:

The Woodpecker

Loves wood boring larvae of beetles, ants, borers and many other insects and their larvae.

The Chickadee

Loves lice, caterpillars, flies and tree hoppers.

The Bluebird

Eats sowbugs, caterpillars and other insects.

Mockingbirds

Will eat termites, caterpillars and beetles.

Robins

Also eat termites, they love caterpillars and beetles.

The Cardinal

Likes grasshoppers, ants and beetles.

Bats

Bats eat a great deal of insects in their life time. Protect them when ever you can.

"Natural Insecticides made from the Earth, Return to the Earth"

It is to our advantage to let Nature be our teacher and help us learn from our mistakes. Better Living Through Chemistry remember that old slogan? has helped us to develop into a highly skilled and educated society. Better Living Through Nature can take us even further by enabling our societies to grow and develop sustainably while we exist in harmony with Nature.

Vegetable Organics

Vegetable Organics is not a new science since house wives have been using garlic or cayenne pepper to control bugs for centuries. Veganics is also a safe method of using not only the wide array of materials available at the grocery store, but also the wide array of vegetables that we can grow ourselves!

Homemade Sprays

Homemade sprays can be made from a variety of plants, herbs, vegetables and various store bought items. A list of the many uses of store bought items can be found in the appendix of this book. They are meant to show you the many ways that you can control pests inside and out without resorting to the heavies such as Raid. Always use with caution. Follow the rule " **Less is Best**". Use as little as you can to achieve the results you want.

A synergistic plant is any type of plant or natural material that when added to another plant/material increases the effectiveness of that plant. Soap is a natural synergist for most plants as it acts both as a wetting agent and pest/disease control. Tabasco sauce is another such product which kills on contact and repels as well. Tabasco soap is a blend of the two.

Many vegetables can be made into a liquid and then sprayed in a diluted form to protect plants ,repell or even kill bugs.

Use all Plant mixtures with Caution

There are many plants which will kill both beneficials and the pests as well, so be specific as to what you are applying or spraying. I have found that allmost all plants can be harmful if improperly introduced into the garden environment.

Tobacco can create havoc in the garden if sprayed with out proper care. Cayenne pepper does the same

damage also. It is for this reason that you must use carefully any sprays that you make.

Also many plants can be harmful to humans, birds, fish. Because it is organic does not mean that it is safe also. One drop of Tobacco Sulfate is enough to kill you. As simple a plant as garlic can do a great deal of damage in the gardens eco-system as it effects the beneficials as well. Many plants are oil based and may produce toxic substances as well. The Oleander plan gives off a toxic substance that could kill a young child or dog. Lettuce when made into a liquid will kill white fly and many other soft bodied insects! Leaves from your tomato plant will kill roaches and flies! Soaps if over used will kill beneficials and earthworms and make for a dead soil. So be careful with these mixtures. Label them and teach yourself, your children how to use them!

Wash off all vegetables, fruits before eating!

Types of Preparations

Infusion

You can make an infusion from almost any plant, vegetable or herb.

An infusion is a fast and simple way to get immediate relief from pest or disease attacks on your plants. This often times will provide pest and disease protection and you will not need to make any thing stronger. Learn to blend different plants to achieve best results.

Pour boiling water on to the leaves, flowers ,roots or seeds placed in a cooking pot or container. Allow to steep for 1 hour. Stiring well. pour thru a strainer. Label for use later. Can be bottled. Store in cool dark place. Use 1/2 to 1 ounce of the material in a pint of water. For best results do not boil. Twigs , stems, roots can also be used in this way, except that you should cut up into small pieces. Always use glass, porcelain or enamel cooking utensils if possible.

Decoction

This is made by simmering the part of the plant,etc., in a nonmetal container, from 1/2 hr up to 2 hrs depending on the material(s) and its use. Always start out with more water then you need to allow for evaporation. Always use filtered water if possible. This method is used for extracting the active ingredients. Store in a dark colored glass. Label for later use. Makes a much more concentrated mixture. This mixture is made when a strong mixture is needed to pro-

NATURAL PEST CONTROL

vide a sure kill or control of the pest or disease in mind. Great care must be taken when using as not to upset the eco-balance. This mixture will kill beneficials as well as pests.

Tincture

A tincture is an even more concentrated extract in a liquid form. It is made thru a process of using a natural liquid that will dissolve the extract into a thick liquid form. You can use alcohol as a base, then evaporate the alcohol leaving behind a strong extract. You can also use citrus oils,vegetable oil, coconut oil, etc., for this purpose.

Wine and other alcohols will also work well. Tinctures are good for making enough of a batch to control a large area or to use on a number of places. Tinctures are also used with inline feeding systems. A 1 ounce tincture can be enough to provide 50 gallons for spraying.

A little goes a long way.

Extract

An extract that is 10 times more concentrated then the tincture. Extracts can be made thru many different processes from steam , high pressure, evaporation by heat, or cold percolation. Here we have usage at the 1 drop per gallon levels. Tabasco sauce is such an extract as is Dr. Bronners Peppermint soap.

Here is a list of some Plants that have been used for centuries for Health, Food and Pest Control.

Most of these you can grow yourself! We are constantly expanding this list with every printing. You can make the preparation suggested or you can obtain from the herbal store.

We Encourage your response:

Anise

To repel pests make an infusion, allow to sit for an

hour add a dash of tabasco and Dr. Bronners Peppermint soap to increase effectiveness. Discourages most chewing insects and caterpillars. Decoctions and Tinctures provide most protection to vegetable plants. Extracts should be used with care as it will repel beneficials also. A good synergist.

Sugar Apple

Use the seeds, leaves and roots to make an extract. Highly toxic to most instaects. Avoid use on vegetables.

Azalea

Dried flowers can be used as a dust, or infusion. A contact and stomach poison.
Avoid use on vegetables.

Balm

When used will repel aphids, ants, and most insects since balm has few pests which like to eat it. Balm oil and Peppermint oil make a powerful repellent.
A good synergist.

Basil or Sweet Basil.

Basil Oil

Is very effective against many insects from mosquito larvae to house flies. An effective synergist to pyrethrum and tobacco. Use 1 oz per gallon for infusion. Experiment. Extracts are very powerful. Add a dash of tabasco and soap.

Beech Family

Effective against borers, and most caterpillars. Use with a little tabasco and soap.

Beet

When mixed with tabasco sauce will effectively repel most flying insects.

Borage

Cocoa butter or coconut oil makes a perfect synergist for this plant. You can also use sun tan oil made from coco butter (like Hawaiian Tropic which is made from mineral oil, coconut oil, cocoa butter, aloe, lanolin, eucalyptus oil, plumeria oil, mango oil, guava oil, papaya oil, passion fruit oil, taro oil and kukui oil). Use only a small amount to start with. Try one tablespoon per gallon. Experiment! These oils will kill many different types of insects both pests and beneficials so be careful when using! Will kill soft-bodied insects.

Make solar tea from dried leaves and flowers (see appendix on preparing solar teas). Add one cup of dried leaves or flowers into the leg of some pantyhose, and place into a one-gallon glass container. Add distilled or filtered or spring water, and one tablespoon melted cocoa butter, coconut oil or the tan lotion mentioned above, per gallon of solar tea. Add one cup Nitron A35(or any liquid seaweed) per gallon. Add one drop natural soap (such as Dr. Bronners Peppermint soap), depending on the insect that it is to be used against. Add a dash of Tobasco sauce to increase effectiveness. Test first for strength.

You can make solar tea out of Peppermint and Spearmint, and herb teas such as Lemon grass, Citrus and Lipton tea. Start by placing 1 cup dried herb into panty hose. Tie into a ball. Place into 1 gallon glass container. Add 10 drops Superseaweed or 1/2 cup seaweed powder per gallon and 1 drop natural soap and a dash of Tobasco sauce. Place in sun for two to three days. Strain. Add to sprayer and use on plants. add a dash of Tobassco sauce to increase effectiveness.

Experiment with various strengths, length on time in sun, etc. Experiment with different types of herbs and keep notes on the effectiveness of your different mixtures. If you find something that works better than anything mentioned here, please send it to me for possible inclusion in the next revision. Wash off all fruits and vegetables before eating!

Cabbage

Leaves can be used to attract aphids to traps. Add a dash of Tabasco soap.

Canna Family

Tuberous rootstocks, broad leaves and showy flowers. The dried leaves and stems contain an insecticide that is as strong as pyrethrum on most insects we tried it on. A dash of tabasco soap will increase effectiveness.

Caraway

Creates strong repellent against chewing insects. An infusion with a dash of tabasco soap will protect most plants. Test for strength.

Castor Oil

The Castor Bean plant. Beans produce Castor Oil, makes a perfect synergist for pyrethrum and other natural insecticides. Make alcohol concentrate from crushed beans or buy concentrated oil. Use 1oz per gallon. Do not use on vegetables or fruits!

Catnip

Prevents insects from establishing on plants.Prevents worms or caterpillars if sprayed regularly.

Cayenne Pepper

Will destroy many insects which are either dusted or sprayed. A good source is Tabasco Sauce. Made from cayenne pepper, vinegar and a dash of salt. Really great. Read the ingredients to make sure that there are no additives or preservatives. All parts of the plant can be used. The powder can be sprinkled around plants. The pepper can be made into an extract. Experiment with different types of peppers. See pepper chart in appendix..

Celeastraceae

The ThunderGod vine has been used as a common insecticidal plant in southern China for many years. The poison comes from the root bark. Alcoholic extracts of the roots are more toxic. Powdered fresh small roots are toxic to first stage larvae of many varieties of moths as well as to cockroaches. About 1/2 as toxic as pyrethrum.

Chamomile

An extract will repel beetles and chewing insects.

Don't Panic It's Organic!

NATURAL PEST CONTROL

Add a dash of tabasco soap.

Chenopodiaceae

Goosefoot Family

Is used as a general insecticide on many different insects through out the world. Avoid using on fruits and vegetables.

The American Chestnut.

A tannic acid extract made from the nut, bark and leaves. Proven to be effective against a variety of insects.

Chives

Use with a natural soap to repel insects.

Citronella Oil

Used for centuries as a mosquito repellent, can also be used as a synergist for many natural sprays. Avoid use on Oil/based Trees(such as pine trees). Natrapel repellent is made from citronella oil and aloe. Repels many insects.

Citrus Oil

Contains Limonene. An extract made from citrus peels.
Makes an excellent synergist to many natural sprays. Will effectively kill many soft body insects and repels many others. Use only on affected areas. Use not more then 10 drops per gallon of the extract, depending on the pest. Use leaves, flowers, grinds. Many companies on the market with citrus products such as soaps. Citrus or Lime juice can be sprayed on most plants and vegetables. Will control fleas, larvae, and most chewing insects. Citrus soap makes an excellent synergist for most extracts. There are many citrus extracts available on the markets today. try some of the air fresheners made from citrus. Use only a few drops per gallon.Citrus Oils are powerful extracts, use with care as it will kill off beneficials as well.

Coconut Oil

Many parts of this palm can be used. From the leaves to its sap. The easiest part to use is what comes from the nut itself. The oil. This oil is found in many suntan lotions. Read the label. Buy only 100% pure. Coconut oil and soap (coconut oil soap is the best) are a great combination. This oil is safe for humans to use but be careful to use only on specific areas as this oil will kill many insects. Can be used as an excellent synergist to be used against hard shelled insects such as snails, beetles, etc. Extracts can be made using a slow heat method.

Give Pests a Coffee Break?

Coffea Arabica L:

Caffeine has several strong chemical compounds that are insecticidal. Soap and Tobasco sauce make an excellent synergist for this plant. Add one to 10 drops per gallon; the strength you use depends on the type of plants you are spraying. Use only natural soaps like Dr. Bronners, Citrus Soap, Herbal soaps, etc. DE also makes an excellent synergist for coffee.

Method #1

Save the unused coffee in a glass gallon container. Use as needed. Can also be fed directly to plants.

Method #2

An effective way to control most insects and also help to provide nutrition is to use your coffee grinds and to make a coffee solar 'brew'. This is done by dumping the coffee grinds from the past few days into a panty hose. Next, place it inside a one-gallon wide-mouth glass container of water and allow to sit for 24 hours in the sun.

The next day, add one cup of Nitron A35(available from Nitron Company), or any natural seaweed product, per gallon. Also add one drop of any of the natural soaps mentioned. Add 10 drops Superseaweed per gallon, and one tablespoonful of any kind of rock dust plus a dash of Tabasco sauce. This mixture will reduce insects and help to control many more.

The mixture should be stirred clockwise for five minutes then counterclockwise for another five minutes before it is strained.

Method #3

Instant Coffee Formula. Add one cup strong instant coffee to a gallon of water. Let sit for 24 hours before adding one drop of bio-degradable soap per gallon and a dash of Tabasco sauce.

Make the mixture as strong as needed, but be careful not to put too much soap in as it could burn the plants. A good rule of thumb for spraying is to spray less. Learn how to use soap. See appendix on soap. Add 1/2 cup seaweed or 10 drops Superseaweed per gallon.

Compositae

Thistle or Aster Family: Chrysanthemum cinerari-

NATURAL PEST CONTROL

aefolium. is the plants name which produces pyrethrum. The flowers are dried and the powder is used. To this family belongs Dahlia, coreopsis, marigolds, aster, cosmos, and many others. Harmless to mammals.

Corn

Corn Oil is an excellent oil to use on various insects and diseases.

Coriander

Makes an excellent repellent. stops chewing insects. Add a dash of tabasco soap. Seeds are used.

Cucumber

Very strong repellent of worms, ants , fleas beetles. Seeds are the strongest.

Dill

An extract made from dill will repel most flying insects. interferes with receiving signals.

Eggplant

A part of the night shade family. Make an infusion for best results. Eggplant oil is good also.

Eucalyptus Oil

Makes an excellent synergist for many natural sprays plus will kill on contact many soft bodied insects as well as repel them.Try Dr. Bronner's Eucalyptus soap.

Fennel

Stops all chewing insects if used as an extract. 1 drop per gallon plus a dash of tabasco soap.

Garlic

You can use all parts of the plant. Use the greens, run through a juicer. Add 1 cup per gallon, allow to sit 1 hr before use. Do not allow to sit longer. Garlic Juice made from garlic cloves is stronger. Extracts are readily available or you can make your own. Use 1 oz per gallon liquid for most insects. Use soap as synergist.

Gourd Family

Gourd melon, squash, and pumpkin all in same family. Seeds are crushed into oil or dust concentrate. Leaves can also be dried and used as an infusion, oil extracts can be also sprayed. Will repel most chewing insects.

Horseradish

A very strong plant. most insects will not eat this plant or have anything to do with it! I suggest getting an extract. from the roots.

Hyssop

Hyssop Flowers has been said to be as strong as pyrethrum flowers. Use an infusion first if not strong enough make into a tincture.

Kutira Plant

Bark and roots can be used to make concentrate. Avoid breathing or absorption through skin. Controls many variety of insects: aphids, whiteflies,leafhoppers, psyllids, thrips, spider mites, snails, slugs, small animals. Use only on affected areas.
The Kutira plant produces a gum which is an excellent synergist for nicotine. Can be grown. if bought make sure it is organically grown. a few good sources are Chinese Herb companies. Use the extract with care.

Larkspur

Powdered roots are toxic to most insects especially caterpillars.

Lettuce leaves

When made into an infusion will repel aphids and whiteflies. add a dash of tabasco soap. Wash before eating.

Marigold Flowers

Use dried leaves for an infusion spray. Add a dash of tabasco soap.

Marjoram

Most herbs can be made into either an infusion or extract and sprayed. Will repel most insects.

Nasturtiums

An extract can be made. Sprayed on plants will repel aphids.

NATURAL PEST CONTROL

Neem Tea Oil

Excellent anti fungus, can repel many insects and destroy others. Use only on affected areas. Will also repel most insects. Oils are easy to find. Avoid use on vegetables.

Nettles

Use both roots and dried leaves as well as flowers, same as above.

Nutmeg Oil

Used as a synergist. Avoid using on fruits and vegetables.

LARGE PARIS ARTICHOKE.

Oleander Oil

Controls many soft bodied insects. Use 1 oz per gallon liquid. Avoid using on fruits and vegetables.

Onion family

Makes a strong concentrate,use dried onions or liquid prep. Just like making soup. Allow to cool. Add a dash of Tobasco sauce . Experiment with different amounts.

Oregano

Make an infusion. Will repel most insects. Add a dash of tabasco soap.

Oxeye

Heliopsis scabra. contains compounds harmful to many varieties of flies and other types of insects. As toxic as pyrethrum against insects such as aphids. Avoid using on fruits and vegetables.

Parsley

When an infusion is made, use right away. Spray on most plants. Add a dash of soap to increase effectiveness.

Peas Leaves

Can be made into an infusion that will repel most chewing insects.

PennyRoyal Oil

An excellent synergist for many natural sprays. Will confuse insects and/or repel. Some insects are destroyed on contact. Make an infusion of the herb.

Peppermint Oil

Makes an excellent synergist for many natural sprays, will confuse many insects. A good source is Dr Bronners Peppermint soap. Contains peppermint oil as a base. This oil will kill, repel or control all types of insects as well as beneficials so be careful. Make an infusion of the herb.

Pigwort

A strong extract, toxic to most insects. Use lightly.

Pine Oil

Kills ants when mixed with 1/2 water. Use only on ants, do not spray plants directly.

Potato

An extract will control beetles, and most chewing insects. Add a dash of the Tabasco Soap.

Quassia

Bark and wood chips contains insecticidal properties. An extract will control many insects. Avoid use on vegetables or fruits. Allow chips to simmer for 4 hours over low heat. Add a dash of tabasco soap to increase effectiveness.

Radish

An infusion of the leaves when added to a dash of tabasco soap will repel most chewing insects, whiteflies, ants and some animals.

Rosemary

Use roots, leaves, flowers, use same as above. An infusion of leaves makes an excellent repellent of many insects. Add a dash of any natural soap.

Rotenone

A better known plant made from the roots of a Tropical Derris Tree and from the cube root. It has been used world wide for many pests for many centuries. Very low toxicity to man and animals. It is a contact stomach poison. It is very effective against the following pests: all soft bodied insects, various spiders, spider mites, flies, snails, fleas, ticks, beetles,

NATURAL PEST CONTROL

leaf roller, borers, etc. Sold in 1% and 5% solutions. Available from Gardens Alive!, Peaceful Valley Farm supplies. Rotenone/Copper mixtures provide disease controls from insects while controlling the insects too. Dangerous to fish.

Ryania

A botanically derived natural insecticide safe to use for mammals, birds, etc. Made from the ryania plant found in Trinidad and is in the same family as tobacco. Ryania affects insects eating and insects starve. Used as an extract will control and repel chewing caterpillars, moths, spider mites, most chewing insects, ants, flea beetles, leafhoppers, cockroaches, aphids, silverfish, spiders, thrips, whiteflies, Japanese beetles and more. Add a dash of any natural soap to increase effectiveness. Can be mixed with other extracts.

Rue

Here is another powerful herb that when made into an extract will provide protect from most chewing insects. Add a dash of natural soap.

Sabadilla

A member of the lilly family. The seeds are highly effective when ground up. A contact and stomach poison. The alkaloids in sabadilla affect the nervous systems of insects. Highly bio-degradable leaving no harmful residues in the environment. To Control aphids, blister beetles, cabbage worms, citrus thrips, german cockroaches, grasshoppers, chinch bugs. Add a dash of Tabasco Soap. Will also kill many beneficials like bees so use carefully.

Sage

The dried leaves of this plant can be used to make a solar tea. Add one cup of dried leaves to a panty hose and place into a gallon glass container filled with filtered, well, or spring water. Allow it to sit in sun for 24 hours before adding one drop of soap per gallon or you can make an infusion of leaves.

Active principles of the plant are the alkaloids contained in the roots and leaves as well as in the seeds. Moderately toxic to most insects but safe to use. A stomach poison for insects causing them to stop feeding after ingesting. Effective in hot weather. Can be combined with pyrethrum and rotenone for max. effectiveness. Controls many harder to control pests such as citrus thrips, corn borers, oriental fruit worm, corn earworm, codling moths.

Sesame Seed Oil

Can be used as a synergist. Use 1oz per gallon liquid used. Try Sesame seed oil and Cayenne pepper mix available in food stores.

Spurge Family

Croton tiglium contains croton oil used in China. The seeds are crushed into oil. Croton resin is more toxic than rotenone to insects.

Tansy

An infusion of this herb with a dash of natural soap will deter most insects.

Teas

Many teas contain tannic acid which deters insects from eating. Use with a dash of natural soap. Try Lipton tea or celestial seasons.

NATURAL PEST CONTROL

Thyme

Another herb that takes well to infusion but extracts are more powerful. Sprayed on plants will control chewing insects.

Tobacco

Nicotiana spp. Tobacco:

Active ingredient Nicotine Alkaloid, Nornicotine.

Sold as Nicotine Sulfate in liquid form. Very Toxic to mammals. A Contact poison and a fumigant. Nicotine Sulfate biodegrades rapidly, acts extremely fast and no insect is immune to it. Be careful when using. Use only small amounts. Use Nicotine Sulfate only if you are a professional and know how to hande deadly chemicals. A safe source is smoking tobacco. Buried at base of plants. See Using Tobacco in appendix. The best to use is Organically grown without chemical addititives. Try Santa Fe Natural Tobacco Company in Santa Fe, New Mexico . They sell both the dried leaves(hands) and ready to smoke. You can buy it by the pound. See resources for address.Tell em we sent ya!

Methods of Application

As a Liquid or as a dust(or powder) or as a concentrate(Black Leaf). Making a solar tea from 1 cup dried leaves is a safer way to use tobacco. Add 1 cup tobacco leaves into a panty hose, tight into knot and suspend into gallon of filtered water. Allow to sit for 24 hrs.Spray on plants or pests. Allow 24 hrs before harvesting. Add 2 tablespoonful of any bio-degradable soap(coconut oil increases effectiveness of tobacco). SunSpray oil can also used at 4 tsp. per gallon of tobacco solar tea made.

Tomato Plant

A member of the night shade family, leaves can be dried, made into a tea and sprayed for many different types of pests. An infusion will control many chewing insects. An extract is more effective.

TGM+ (Tobacco, Garlic, Manure and Rock Dust)

Can also be buried under plants. Plants absorb nicotine/garlic which kills any pests attacking it. Not harmful to plants, trees, roses, ornamentals. Do not use on fruit tree during fruiting stages.

See index under tree vents for info on how to use.

Wild flowers

Most wild flowers can be made into an infusion or an extract. Experiment.

CALLERY PEAR

Wormwood

A powerful when made into an extract. Be careful with its use. Controls most chewing insects,snails and slugs.

Using Soap

Pure soap is a common ingredient found in the organic Gardeners list of safe sprays to use. Most oils come from fats and oils found in animals,(Vegetable and plant derivations are more effective and environmentally sound then from petroleum based) and some come from plants and others from trees(coconut). It is useful against many insects as well as being a wetting agent. The most famous soap is Safers Insecticidal soap which can be found in many nurseries and available from Gardens Alive, Nitron. Safer soap is made from fatty acids found in animals. See Soap in index for more references. Dr. Bronner's Soaps have been around for a long time also.

Safer Soap

Controls most soft bodied insects, the soap must come into contact and therefore called a contact spray. Safer Insecticidal soap also comes with pyrethrum and contains no Piperonyl Butoxide. Controls aphids, thrips, mealybugs, spider mites, whitefly. Safer soap also comes with citrus aromatics.

Another excellent soap to use is Shaklees Basic H formula available through a Shaklee distributor. Shaklees Basic H will kill most insects on contact. Since it is a concentrate use small test amounts first. Completely safe and biodegradable.

My favorite soap to use is Dr. Bronners Peppermint soap, available at most health food stores. A natural soap made from fatty acids and peppermint oil.

NATURAL PEST CONTROL

When using soap be careful not to use too much as soap will burn the plants, it is always wise to test the soap out on a small section of the plant to make sure it won't harm the plants.

"When in doubt use less!"

Hints on buying soap

It is wise not to buy soaps that have a variety of unnatural additives, colorings and other 'inert' ingredients. Do not use dish washing liquids! They are worse then most chemicals!

Here is a list of some of the more well known brands of safe organic soaps available which can be converted for our use as a natural part of our IPM program:

Safer Insecticidal soaps, Pure Castile soap, Vegetable glycerin soap, Pure Coconut oil soap, Pure Olive Oil soap, Dr. Bronners soaps, Pure Herbal soaps, Boraxo Pure soap, Shaklee's Basic H, Amway's LOC, Citrus Soap.

See Resource directory for addresses.

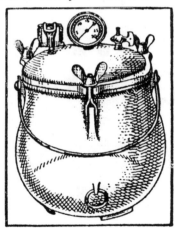

Do-it-Yourself

You can make your own Natural liquid soap:
Boil 1 gallon water.
Stir in 1 cup instant Lipton tea.
Stir in 1 cup instant coffee.
Stir in 1 cup dia-earth.

Add 5 drops of any natural scent that you like such as peppermint, or citronella oil. Add soap flakes made from your favorite natural bar soap. Heat over low heat till boiling. Lower heat and simmer for 15 minutes. Pour into containers. Allow to cool.

This mixture can be used when ever you need to

spray soap. Slice a strip of the soap and add to water and dissolve. Use apprx. enough soap to effectively either kill the pest or effectively repel it. Remember, that too much soap will also damage the plants and the soil, so determine the correct amount that you will need for the job. Use a small amount on a test plant then record your results for later use. Nutritional deficiencies are generally the causes for the various problems that attack plants.

Tabasco Soap

Making your own Tabasco Soap

Purchase your favorite natural soap. I suggest Dr. Bronners Peppermint soap as the most effective to use and also get a bottle of Tabasco sauce(without additives etc.) A good mixture is 8 oz of Dr. Bronners Peppermint to 1 oz Tabasco sauce. This mixture will tend to separate so stir well before using. Label it and store in cool place. Experiment with other soaps.

Using Dormant Horticultural Oils

Horticultural oils kill by suffocating insects and their eggs. Beneficial parasites and predators are not affected by this oil. This brand horticultural oil,can be used on serious pests in the garden without upsetting the balance. Also helps to control many different types of fungus when sprayed during dormant period.

Ultra Fine Oil

SunSpray is an Ultra-fine Horticultural spray oil sold by Gardens Alive! This oil provides increased protection from phototoxicity (the burning of plants leaves). This oil can also be used year round. Can be used up to day of harvest as well as year round! Also acts as a repellent keeping pests away for days after spraying. Ultra-fine oil is available from Gardens Alive!, Peaceful Valley, ARBICO, Garden-Ville,etc. Controls aphids, mites, beetle, leafminer, leafhopper, thrips, whiteflies, scale, psylla, tent caterpillar, borers, mealybug and more! Ok for vegetables and fruits, ornamentals, berries.

If you have the nutritional aspects covered and a problem still arises then to determine what the problem is and what the proper spraying formula should be follow these few steps:

1.. avoid spraying, use as a last recourse.
2....locate stress factors
3....isolate the stress factors and correct it.
4.....monitor closely
5.....talk to others about it.

See **Pest Control Chart for more information** (next page)..

NATURAL PEST CONTROL CHART

Legend: ❋ = Excellent to use   ● = Good Results

Pest	Hand Pick	Traps	Ant Cafe(TM)	Snail Inn(TM)	Stick'ems	Shaklee's Basic H	Dr. Bronners Peppermint Soap	Insecticidal Soap	Water-Wash Off	Pyrethrum	Diatomaceous Earth	Boric Acid	Milky Spore Disease	Fine Horticultural Oil	Beneficial Nematodes	Citronella Oil (repels)	Rotenone	Cayenne Pepper	Tobasco Sauce	MVP/BT	Garlic Spray	Vitamin D	Tobacco Leaves
Ants			●			❋	❋			●	●	●	❋			●	❋		❋	●		●	●
Aphids	●				●	❋	●	❋	●	●	●					❋	●	❋	●				❋
Beetles					❋	●	●	●		●	●					●	❋	●		●	❋	●	●
Borers	●					●	●			●						●		●		●	❋	●	❋
Caterpillers	●																●	●	●	❋			❋
Cockroaches		❋			❋	❋	❋			❋	❋	❋				❋	●	❋	●		❋		●
Fleas		❋				❋	❋			❋	❋	❋				❋	●	❋			❋		●
Flys		❋			❋	●	❋				●	❋	●			❋	●	❋	●				●
Gophers/Moles		❋		Plant Gopher Purge					●							❋		❋			❋	❋	
Grasshoppers	●	❋			❋	●	❋	●	●	●	●	●				●	❋	●	●	❋	❋		●
Grubs	●											●		❋		●					❋		●
Japanese Beetle		❋				●	●	●				●		❋		❋	❋	●			❋		●
Lawn Moths		❋			●	❋	❋	❋			●	●				❋	●	●		❋	●		❋
LeafHoppers	●																●	●	●		●		❋
Leaf Miners	●																●	●	●				❋
Mealybugs	●	❋				❋	❋	❋	❋	❋	●	●	●		❋	●	❋	●	●	❋		❋	❋
Nematodes															❋								
Rodents		❋			❋												❋		●		●	❋	●
Scales	●						●	●	●	●				❋			●	●	❋				●
Spiders		❋			❋	❋	❋	●	❋	●	●					❋	●	❋		❋		❋
Spider Mites		❋			●	❋	❋	❋	❋	❋	❋	❋	●			❋	●	●	❋		❋		❋
Snails/Slugs	●	❋		❋	❋	❋				❋	●							●	❋				
Thrips					❋	●	●	●		●	●				❋		❋	●	●	❋		❋	❋
Whiteflies		❋			❋	❋	❋	❋	❋	❋	●				❋		❋	●	●	❋		❋	❋

Always provide nutrition, use SuperSeaWeed, etc.

Chapter 6
Natural Disease Control

There are many diseases found in the garden. All diseases are due to any one of these factors or a combination of each:

Dead Soil due to over Chemicalization.

High Nitrogen Fertilizers.

Poor Health of Plants.

Stressed Out Plants.

Water Soaked Soil.

Some Diseases Introduced by Viruses.

Some Diseases Introduced by Weeds.

Some Diseases Spread by Insects.

Improper Varieties Planted.

Improper Planting Time.

Improper Watering Techniques.

How Important is it to Identify the Disease?

Diseases in different plants may have the same name but will show different symptoms thus making Identifying very difficult if not impossible. While I encourage you to try and Identify the disease that is attacking your plant(s), you must remember that the basic treatment(s) remains the same regardless of the disease.

A Few Points to Remember

When dealing with diseases, it's the cause which you must control, or cure, rather then the affect (disease).

If you Eliminate the Cause, you Eliminate it's Effects.

Always key into Nutrition First.

Lack of nutrition promotes stress. Chemical fertilizers do not make complete food sources for plants. They cause imbalance both in the plants and in the soil.

The Greater the Stress
The Greater the Disease

Always Seek Balance

When things are out of balance, infestation/disease occurs.

The Good Guys and the Bad Guys don't live in the same place.

Bad bacteria and good bacteria do not share the same soil. Dead soil is actually not dead but instead contains bad bacteria or fungi or any number of disease sources.

I will not attempt to describe each type of disease for you in this book. There are many books out that which do just that, instead I will point you in the right direction since it has been my experience that you need not know the name of the disease to treat it..

Remember its either a pest or a disease. All diseases are treated basically the same way. as long as you treat the cause, the effects will disappear. If the disease is persistent after following the below steps then proceed on to the organic disease control section.

Please see appendix for suggested books for more information..

Steps to take in Controlling Plant Diseases

1...Proper placement of the garden is important. Choose a location which gets as much light as possible. Do not plant under or near a tree or next to a lawn.

2...Amend the soil with good rich alive compost. Apply mulch as needed. You should have plenty of earthworms. Dead soils have 0 earthworms.

NATURAL PEST CONTROL

Natural Disease Controls

Minerals

3...Choose the correct variety for your area. Heirloom varieties are stronger and more suited for organic gardening. Write to Abundant Life Foundation or Seeds of Change. See Resources Directory for their address.

4..Practice soil rotation if possible. Rotation of crops is suggested whenever possible. If a raised bed is used allow the bed to go fallow for one season. Amend with compost, rock dust and animal manure if available.

5...Use disease free seedlings and plants. Many diseases are carried into the garden. Be especially careful when bring potted plants with soil into the garden.

6...The stronger the plant the less the disease will affect it. Avoid using any chemicals in the garden. Avoid High nitrogen fertilizers.

7...Certain weeds are hosts to pests which carry the disease into your garden, therefore weed control of some type is important. Weeds also provide homes for many beneficals so choose carefully.

8...Use Trap crops in addition to various types of traps which attracts the pests away from the garden.

9...Destroy diseased plants. Throw away in the trash. Do not compost since most backyard gardeners do not compost in big enough piles in which the temperature is high enough to kill of the fungi.

10..Avoid spreading the disease by cleaning your tools with alcohol inbetween cuttings..

11..Avoid overhead watering. Use a soaker hose or a drip system when ever possible.

12..Do not use chemical fungicides. They only will cause more problems both to your health and the soils health.

13..Use a Garden Filter when ever possible. Chlorine will kill the bacteria in the soil and this will cause problems. See Garden Filter in appendix.

If all of the above fail to control your disease(s) then there are many natural disease controls which you can resort to.

There are over 72 minerals which are needed by the soil and the plants for healthy disease free growth. Here are just a few which you can work with:

Bordeaux

Is a mixture of copper sulfate and lime. Used in France for at least 100 years. Very poisonous, be careful applying. Follow instructions on label. Good for fungus diseases on grapes, fruits such as peaches, apples. Used against Potato blight. Helps roses with fungus. Available from Gardens Alive! Peaceful Valley, ARBICO, etc and at most nurseries.

Calcium

Calcium can be used to reduce certain types of bacteria, fungi found in certain diseases. Also beneficial bacteria will enjoy the calcium in their diet. So a good spray will naturally be a mixture of calcium and a bacterial source. Calcium can be obtained from crushed egg shells. Bone meal is also a good source of calcium. Milk provides calcium and a beneficial bacteria(see milk). Rock dust is also high in calcium(see rock dust).

Copper

Copper is an excellent natural fungicide, controls diseases of vine crops, potatoes, leaf spot, anthracnose, downy mildew, powdery mildew, scab, fire blight, bacterial spot. Available in liquid or dust form. Available from Gardens Alive, ARBICO, Peaceful Valley, and most nurseries. Use sparingly. Prevents black spot on roses, tomatoes; and other diseases. A rotenone/copper mixture will also control insects. Excellent for use on melons and cucumbers. Can be used as a liquid for spraying or as a dust for dusting.

Magnesium

Is available in Epson salts. available at most drug stores. Magnesium will help to reduce stress on the plants. Plants can become stressed, out due to minerals not being available to them. High nitrogen causes imbalances in the soil, and plants bio-mechanisms, resulting in the locking up of trace minerals in the soil, and the in ability of the plant to intake the required minerals.

Sulfur

Is an ancient tool of the organic farmer. A natural fungicide and insecticide. Used for rust, powdery mildew, leaf spot, brown rot. Also a good soil acidifier. Available from Peaceful Valley, Gardens Alive! , ARBICO, most nurseries. Micronized sulfur is an extremely fine dust which when dusted on plants will control apple scab, cedar-apple rust, black rot, leaf spot, and powdery mildew. Liquid sulfur is also available.

Zinc

This exotic element is just as important as the top three(NPK) without which the plant would die. Zinc is normally readily available but when the flow of the over 72 minerals needed by the soil and plants is interrupted, the zinc will be used up by the plants. When the plant don't even have any zinc left they go into immediate stress. Good rich compost will provide the zinc.

Natural Sprays

Alfalfa Tea

A tea can be made using Alfalfa meal. I suggest that you add 1 cup per gallon of the meal. Allow to sit for 1 hour, strain thru a gardeners cheesecloth. Spray on plants to control most fungus before they start. Add a dash of natural soap as a wetting agent. Great as a blend with any of the minerals mentioned above.

Baking Soda

Baking soda can be made into a spray at the rate of 1-5 tablespoons per gallon(depending on the plant sprayed). Baking soda and milk with a dash of natural soap makes an excellent fungal control. See chart for list of diseases it controls. A better mixture is baking soda, dash of soap and manure tea.

BD 508

A biodynamic spray which will help to prevent many fungal diseases. See Chart. Available from the Bio-Dynamic foundation. See resource directory for address.

Clay

Clay can be used as a fungal control. Add 1 cup of finely powdered clay into a gallon of fil-

tered water, allow to sit for an hour. Strain thru gardeners cheesecloth. Add seaweed to increase effectiveness. Use either red or white clay.

Compost Tea

Compost tea when properly made will control many fungal diseases. Place 1 cup of your compost into a panty hose. Tie into a ball. allow to sit in filtered water. The length of time depends on the severity of the problem. Allow to sit for apprx 1-5 hours. Add a dash of natural soap. Then spray on plants.

Fish Emulsion

Fish emulsion will provide essential oils that will prevent many diseases from occurring. Use 5 tablespoons per gallon. Avoid Urea based products. High nitrogen is not good for disease control.

Garlic

Has been used for centuries as a fungal control.Garlic can be made into an extract and sprayed. Add a dash of natural soap to increase effectiveness. Garlic can also be grown around the plants. The natural aroma will provide for a fungus free environment. This is mainly due to the garlic being absorbed into the plant.

Horsetail

An old time favorite of Bio-Dynamic gardeners. Effective against many bacterial diseases. An infusion of the leaves, flowers can be used for most problems., for greater strength make an extract or obtain thru mail order companies. The dried herb is availbe thru many mail order herb companies. Also available thru Peaceful Valley Farm supply. For better results mix with a dash of soap and a dash of garlic.

Hydrogen Peroxide

This product will reduce many types of fungal diseases. Use only as a temp control. Use 1-5 tablespoons per gallon water of the 3% (best to use food grade at 1 drop per gallon). Always test for strength as not to burn plants. Always use filtered water when making any natural bacterial sprays. Use seaweed or natural soap to increase effectiveness.

Lime

Lime will control many diseases around the garden. Use with care as it will kill off benificials as well.

Milk

The high calcium levels combined with

NATURAL PEST CONTROL

milks natural bacteria make for an excellent natural spray to control fungal diseases on many plants. Add a dash of garlic and a dash of baking soda. Raw milk is best but otherwise it doesn't make a difference the amount of fat it has. Powdered milk will work if its only the calcium you want.

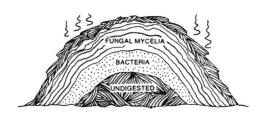

Manure Tea

Manure tea makes an excellent anti-fungal spray as it introduces many different types of beneficial bacteria. Avoid using raw manures. Use only well aged. See manure tea in appendix for preparation. Add a dash of natural soap to increase effectiveness. Manure can be placed into panty hose , tied into a ball and allowed to sit in filtered water.

adult moth

Molasses

Will instantly raise the energy level of the plant. This instant raise will help to deter many fungal diseases and pests. Keeping pests away is a very important part of controlling diseases naturally. Add a dash of either natural soap or fish emulsion to increase effectiveness. Goes well with milk and also with rock dust. A good strong anti fungul formula is mix equal amounts of molasses, seaweed powder, milk and rock dust(about 1 cup ea), mix well until a paste. You will need to add more rock dust and seaweed until paste. You can then place small amount of the paste into a panty hose and tie into a ball,. Allow to sit for 2-4 hrs. Then spray on diseased plants.

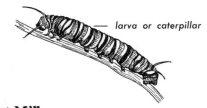

— larva or caterpillar

Rock Dust Milk

Rock dust has very large amounts of calcium, iron, magnesium as well many minerals which make it a quick energy boast to plants and increasing the beneficial bacteria count in the soil. Rock dust can be added to filtered water and turned into a milky liquid and sprayed or used as dust. I suggest adding a dash of natural soap along with a dash of seaweed. A dash would be 1 tablespoon of powdered seaweed or 1 ounce of liquid seaweed per gallon of filtered water. See rock dust in appendix for its preparation.

SunSpray Ultra Fine Horticultural Oil

Can be used to control many fungal diseases. Sprayed year round will prevent diseases from spreading. Great on fruits and vegetables, roses and most flowers. Avoid on hairy plants. Avoid if hot days(over 90 degrees). Adding a dash of garlic and natural soap will increase it's effectiveness.

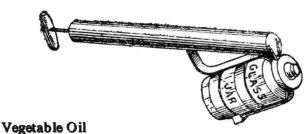

Vegetable Oil

Can be used as a way of controlling many diseases when they first occur. Should be used during warm or cool weather. Avoid using during hot weather. Corn oil, soy bean oil, olive oil, coconut oil are but a few oils you can use. See Oils chart.

DISEASE CONTROL CHART

Good Results ●
Excellent to use! ✱

Disease	Compost Tea	SuperSeaWeed	Milk	Copper	Liquid Sulfur	Bordeaux	Liquid Garlic Spray	Lime/Sulfur	Fine Horticultural Oil	Garlic Oil	Vegetable Oil	Shure Crop	Manure Tea	Liquid Seaweed	Nitron A-35	Soapy Water	Fish Emulsion	Rock Dust Milk	Molasses	Hydrogen Peroxide	Baking Soda	BD 508	Clay Spray
Anthracnose	✱	✱	●	✱		✱	●	●	●	✱	●	●	●	●	✱	●	●	●	●	✱	●	✱	●
Apple Scab	✱	✱	●	✱		●	●	●		✱	●	●	●	●	✱	●	●	●	●	✱	●	✱	●
Bean Rust	✱	✱	●	✱	●	✱	●	●		✱	●	●	●	●	✱	●	●	●	●	✱	●	✱	●
Gall Formers	✱	✱	●				●	●		✱	●	●		●	✱	●	●	●	●	✱	●	✱	●
Bacterial Blight	✱	✱	●	✱	✱	●	●	●	●	✱	●	●	●	●	✱	●	●	●	●	✱	●	✱	●
Bacterial Canker	✱	✱	●			●	●	●		✱	●	●	●	●	✱	●	●	●	●	✱	●	✱	●
Bacteria Spot	✱	✱	●	●	●	●	●	●	●	✱	●	●	●	●	✱	●	●	●	●	✱	●	✱	●
Bacterial Wilt	✱	✱	●		●	●	●	●		✱	●	●	●	●	✱	●	●	●	●	✱	●	✱	●
Blossom Drop	✱	✱	●								●	✱	●	✱	●	✱		●	●	●		✱	
Black Rot	✱	✱	●	●	●	✱	●	●	●	✱	●	●	●	●	✱	●	●	●	●	✱	●	✱	●
Common Mosaic	✱	✱	●			●	●	●	●	✱	●	●	●	●	✱	●	●	●	●	✱	●	✱	●
Damping Off	✱	✱	●		●		●	●		✱	●	●	✱	●	✱	●	●	●	●	✱	●	✱	●
Downy Mildew	✱	✱	●	✱	✱	✱	●	●	●	✱	●	●	✱	●	✱	●	●	●	●	✱	●	✱	●
Fire Blight	✱	✱	●		✱	✱	✱	●	●	✱	●	●	●	●	✱	●	●	●	●	✱	●	✱	●
Early Blight	✱	✱	●	✱	✱	✱	●	●	●	✱	●	●	●	●	✱	●	●	●	●	✱	●	✱	●
Fusarium Wilt	✱	✱	●	●	●	✱	●	●	●	✱	●	●	●	●	✱	●	●	●	●	✱	●	✱	●
Powdery Mildew	✱	✱	●	✱	✱	✱	●	●	●	✱	●	●	✱	●	✱	●	●	●	●	✱	●	✱	●
Orange Rust	✱	✱	●	✱	✱	✱	●	●	●	✱	●	●	●	●	✱	●	●	●	●	✱	●	✱	●
Rhizoctonia	✱	✱	●	●	●		●	●	●	✱	●	●	●	●	✱	●	●	●	●	✱	●	✱	●
Scab	✱	✱	●	✱		●	●	●	●	✱	●	●	●	●	✱	●	●	●	●	✱	●	✱	✱
Sunscald	✱	✱	●							●			●		✱		●	●	●	✱	●	✱	✱
Tobacco Mosaic	✱	✱	●			●	●	●		✱	●	●	●	●	✱	●	●	●	●	✱	●	✱	●
Verticillium Wilt	✱	✱	●			●				✱	●	●	●	●	✱	●	●	●	●	✱	●	✱	●
Leaf Spot	✱	✱	●	✱		●	●		✱	●	●	●	●	✱	●	●	●	●	✱	●	✱	●	

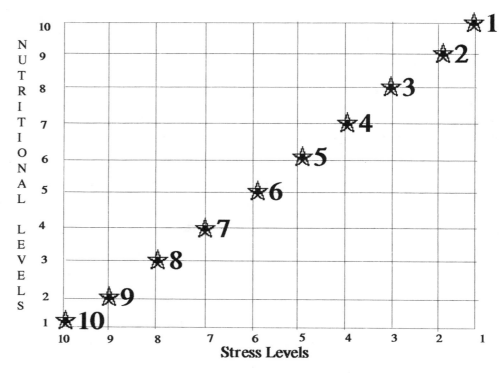

NATURAL PEST CONTROL

"Nutrition is very important in controlling pests"

The chart on this page shows the various pest levels compared to stress and nutrition. High stress levels and low nutrition levels produce high pest levels. Here are some examples:

Pest Level #1 shows a very high nutrition level combined with very low stress level to produce a very low pest level. At this level there are no pest problems.

Pest level #3 shows the nutrition level to have dropped and the stress level to have increased consequently the pest level has increased. At this level pest activity is becoming noticeable. There is still adequate time to control this situation.

Pest level #6 shows a dangerously low nutrition level with a dangerously high stress level resulting in a high pest level. At this level considerable damage occurs. This will take longer to control.

Pest level #8 shows a situation where nutrition is so low as to be almost non-existent with correspondingly high stress levels. At this level the soil is dead or nearly dead and the root systems non-functioning. Foliar feeding as well as immediate soil composting is required. It will take time for results.

Pest level #10 is a disaster both to plants and the soil. Incredibly it is at this level that most chemical farmers operate. With a dead soil eco-system, plants rely solely on the "hit" they get from the chemical fertilizer, and farmers must rely entirely on chemical pesticides to protect their crops. The organic system will not work here!

PEST LEVELS

Nutritional Levels (vertical axis, 1–10)
Stress Levels (horizontal axis, 10–1)

Data points: ☆1 (10,1), ☆2 (9,2), ☆3 (8,3), ☆4 (7,4), ☆5 (6,5), ☆6 (5,6), ☆7 (4,7), ☆8 (3,8), ☆9 (2,9), ☆10 (1,10)

NATURAL PEST CONTROL

Chapter 7
Organic Flea Control

The War against FLEAS.

Lets talk about the cycles of the flea. There are three different types of fleas, the Cat Flea(Ctenocephalides), The Dog Flea(Ctenocephalides) and The Human Flea(Pulex irritans). The Cat Flea is the one we are most interested in as the other two are very rare. Cat Fleas not only like cats, dogs, and humans but they also like many other species such as rats, chickens, etc.

The Life Cycle of the Cat Flea

Adult fleas lay eggs outside in the grass or any warm safe place (usually near a warm blooded host). The eggs develop into legless larva which feeds on tiny amounts of organic matter for up to one month. Usually the egges are laid on the hosts which then drop to the ground. If inside they fall on the carpet, etc. the larvae do not bite but live off dried blood defecated by the adult fleas.Outdoors the larva will lie in a shady moist spot. The larva then becomes a pupae by spining a cocoon and emerge in about a week as a hungry adult.

The cycle for reproduction is 30 to 60 days and is triggered by the presence of the hosts warmth and the carbon dioxide given off by the hosts respiration. The Larvae can live up to 6 months and the [pupae can live to a year until the conditions are right for emergance.

What this means is that if you have just moved into a new apartment or home which had a dog or cat as tenants is that your presence,your cat or dog could trigger off their emergance! Many people have allergic reactions to flea bites while others never notice it. Fleas usually do not bite humans unless their are no other hosts around.

> Regular vacuuming.
> Regular dusting before and after vacuuming with pyrethrum.
> Proper animal grooming and flea combing.
> Bath your dog with Dr. Bronner's Peppermint soap(or any other peppermint base natural soap).
> Proper animal feeding.
> Proper care of animal bedding areas(dustings, vacuuming).
> Proper care of soil and environment (planting herbs).

Control Methods

Your treatment should be spread out over a 60 day period. This means that you should do regular flea treatments every month if you are in areas prone to having fleas.

I have 2 cats and one dog. I know fleas. The cats clean themselves and generally don't like to be told what to do or to have anything done to them but they can be helped as well as the dogs. Fleas and dogs, fleas and cats, they go together. Its un-natural for there never to be fleas again. WE can't totally rid ourselves of them any more then we can totally rid ourselves of the ants. The answer here is to control and reduce them. The formulas given here along with the hints are to help you keep fleas out of the house and off your animals. Follow these hints and you will succeed.

Many flea collars on the market contain Sevin which attacks the nervous system of the insect(as well as your pets nervous system and ourselves as well). I have seen flea collars which contain DDT! The companies are getting smart and are using Pyrethrum but are screwing up when they add piperonyl butoxide to booster it up(see making your own pest control chapter for more info). This product has been associated with liver disorders. You should avoid using this product. Especially avoid using around children or seniors or those with health problems.

Insist on using only pure pyrethrum flowers(the whole plant). Read the ingredients. ARBICO, Earth Herbs, Eco Source, EcoLabs, Gardens Alive!, Nitron Inc.,Garden-Ville and Gardeners Supply are a few places you can buy pure Pyrethrum(see resources directory). Also available at many garden centers, etc.,

NATURAL PEST CONTROL

Steps to Organic Flea Control

Step 1

The first tihing you have to do is to reduce their population asap. This is done by spraying the infested areas with a natural soap that will kill the fleas(but not the eggs). This can be sprayed around pet beddings, rugs, furniture, etc..

Using Natural Soap Sprays

Dr. Bronner's Peppermint Soap

This is an excellent soap to use as a mist around infested areas. Smells great too! See below for bathing your dog instructions.

Shaklees Basic H

A good subsitute if you don't want any scent, yet effective. Use 5 tablespoons per gallon.

Safer Flea and Tick Spray

Pyrethrins and insecticidal soap are combined to make an effective spray. Add the Dr. Bronners to make a powerfull mixture. Add equal amounts of each or you can experiment for strengths. Spray around infested areas. Safer also makes an Insecticidal Flea soap for dogs to bathe them in. Add a dash of the Dr. Bronners to increase effectiveness.

Citrus Based Soaps

Make excellent soaps for controlling fleas.

Your own Liquid Pyrethrum Formula.

1/2 cup Pyrethrum dust, 1 oz alcohol bio-degradable soap

First make a Pyrethrum Slurry by adding 1/2 cup pyrethrum dust to an empty container. Add a small amount of alcohol (1/2 oz). Add a few drops of a bio-degradable soap(such as Shaklees Basic H or Amyay's LOC or Dr Bronners). Slowly add 1/4 cup water, stirring it up until it is all dissolved and forms a slurry. Slowly add the Pyrethrum slurry to a panty

hose, tieing into a ball. This 'ball' is placed in a gallon of water, stirring the water as you go.

When it is all dissolved you can use it to fine mist the rug and the dogs when ever there is any signs of fleas. You can lightly mist the dog before entering the house. You can also spray the area around the house once or twice per week as needed, as well as spraying the dogs bedding area.

Step 2

Using Natural Dusts

There are several different dusting formulas which you can use on fleas , this will depend on your preference. You dust lightly before and after you vacuum.

A Good Natural Flea Control Formula with Pyrethrum

Pure Pyrethrum powder is one of the safest ways to kill and control fleas both on and off animals as it is totally harmless to human or animals. Avoid breathing. Comes from the Pyrethrum plant. Dust dogs once or twice per week, for a bad case use daily. Dust dog bedding and dust rugs, carpets etc before vacuuming. A good approach is to start in one corner of the room and dust slowly as you walk out. Allow to settle for 1 hr before vacuuming.

Nitron sells a mixture of Pyrethrum and Diatomaceous Earth (garden grade) called Diacide. Also Try Gardeners Supply, Peaceful Valley, Gardens Alive!

DE and Salt Formula for Fleas.

1 lb...Dia-Earth 10 oz...Salt

Inside

Mix Dia-earth with the salt and lightly dust the rugs. You can use a salt shaker to dust with or make your own. Allow to sit for a 1 hour then vacuum. You can lightly dust afterwards and leave it on. This will prevent further infestation. Dogs can be dusted with pure dia-earth. dust once per week. Use a small handful and rub on coat. Be careful to avoid the eyes.

Outside

Lightly dust outside areas once per month, using only the DE.

NATURAL PEST CONTROL

Dia-Earth/Pyrethrum Formula.

10 lbs Dia-Earth 1 lb Pyrethrum

Mix the dia-earth with the pyrethrum powder. This is a very strong combination and will kill and control many other insects as well as fleas. Can be used inside or outside.

Diatomaceous Earth is available at: ARBICO, Garden Ville, Peaceful Valley, National Research and Chemical company, Nitron Industries, Universal Diatoms Inc., Gardeners Supply and many mail order companies.

Using Dried Pennyroyal

Sprinkle dried Pennyroyal in rugs, animal bedding, etc. Pennyroyal can also be made into a tea and used as a bath for dogs. Grow more herbs in your environment. Growing herbs such as pennyroyal, peppermint, will naturally repel fleas, etc. Pennyroyal is a fragrant low growing ground cover. Easy to grow and effective against fleas.

You can alternate between any of the above mentioned dusts. Don't use any thing that doesn't feel right for you. Allways test a small portion of the rug to avoid coloring damage.

Step 3

Vacuum, Vacuum, Vacuum

The treatment is regular vacuuming of rugs, carpets and upholstery, pillows, mattresses. etc. Be sure to vacuum the pets sleeping areas as well.

Vacuum at least twice per week, daily is better for bad flea cases. Be sure to properly dispose of the vacuum bag otherwise the fleas will get out(tape close). Vacuuming helps not only to rid yourself of the adult fleas but the eggs and larva as well. I realize that this is a lot of work but neccessary for proper control of fleas.

When fleas are reduced then you can vacuum once or twice per week or as often as neccessary. I am not saying that the answer to fleas is vacuuming! tho it does help. Here are some formulas for killing and controlling fleas both in the home, outside, and on your pet.

Step 4

Using a Flea Comb

There are special flea combs available that you can use to comb your cat or dog. This is a very effective method of controlling and reducing fleas. Monitors fleas as well as controls problem by picking up Flea eggs. You dip the flea comb into soapy water or you can add a little alcohol, to kill the fleas. Available in plastic or steel.

Feeding a little garlic to your pet will help them rid themselves of fleas, and it will also control worms too. Feeding brewers yeast to your animals will also help control fleas on them.

A herbal flea collar would be very usefull here.

Step 5

A Safe Shampoo for Dogs.

Bathing your dog regulary with a natural based soap will go a long way towards reducing pests on your animals. Always read the ingridents before buying. Avoid using any soaps with additives or coloring, dyes and in general avoid using if you can't pronounce the ingredients! Safer Soap shampoo is a safe non chemical shampoo for your dog. It is very effective and mild on the skin. Most nurseries sell it. Be careful using shampoos that use pyrethrum with Piperonyl Butoxide. It's not the Pyrethrum that you should avoid but the additive(see info on Pyrethrum additive chapter "making your own pest controls"). This stuff is harmful to your dog and you as well. A good safe shampoo is called Flea Stop which contains D-Limonene (comes from lemon peels) which kills fleas dead.

NATURAL PEST CONTROL

Dr. Bronner's Peppermint Soap

Bathing your dog in Dr. Bronners Peppermint soap will kill all fleas and ticks while making your dog smell nice. You can also add a teaspoon of Dr. Bronners into a quart sprayer filled with water and use to spray or mist your dog before coming into the house. Use a towel to dry off. This mixture can also be sprayed around the dogs bedding area and around the entrances to the house as mentioned in step 1.

Safer Insecticidal Soap

Controls fleas without the chemicals! Add a dash of Dr. Bronners Peppermint soap to increase effectiviness.

Tea Any One?

Next time you make peppermint tea, save the tea bag. Next time you bathe your dog, make a batch of pepermint tea(using the old tea bag you saved). Allow to cool and use as a rinse on your dog.

Step 6

Birth Control for Fleas.

Continual use of the dusts, sprays and vacuuming mentioned above will provide a form of birth control for fleas. I suggest dusting once per month during flea season and spraying as often as needed.

Step 7

Here Fleas, Here Fleas A Flea Trap

Fleas like most insects are attracted to heat and light.This trap uses a light bulb and a sticky sweet smelling non-toxic, adhesive paper disc that lines the bottom.. This trap works 24 hrs, day and night inside any room. The fleas in the carpets are attracted to the light and heat given off by the bulb and are stuck on the sticky mat. There are many compnaies now selling these ready made. Try Gardens Alive! or Shaper Image. A simple home made version would be using a light clamp, shining the light onto a sticky mat. You can also use a bowel of soapy water and shine the light on the water or you can use cheap wine. The wine will also attract them. The light shined on the wine will make the wine scent travel farther.

BIO FLEA HALT!

Is relativly new nematode product which controls fleas outside! Using specfic beneficial nematodes to control flea larvae and pupae. It will effectively kill over 90% of both the larvae and pupae within 24 hrs. Safe to use around the vegetable garden , lawns, etc. Contact Farnam Companies of Phoenix Az for more info on local sources.

Some Helpful Hints

For long term control of fleas on your property, you must strive to constantly raise the energy level of your property. This is done by providing your soil with plenty of good rich compost, by regular mulching and by promoting a biological diversity. It will be this diversity that will automatically control the infestation of any one insect.

"One of the best controls for fleas in a healthy animal is proper nutrition and exercise"

Stress can make an animal or human more prone to flea attacks. Reducing stress will reduce pest attacks. The principle laws of balance are the same in all living things. From the smallest to the largest, all living things require balanced energy. The equation of life dictates that harmony be Balanced by evolution or growth. In animals, this balance is maintained thru proper nutrition, proper elimination and proper exercise.

Keep yourself healthy by eating and exercising. Keep your pets in shape by feeding them good food and giving them plenty of exercise. Read the ingrients of the food you are feeding your animals. Avoid foods with chemicals as additives.

Remember, Happy animals makes for healthy living!

"Why save the dolphins? In the process we might just save ourselves. My belief is that the important lesson to be learned is in the evolution of caring about ourselves as individuals to the altruistic caring for the larger group. Coming to this understanding, may be the key to our survival as one of the many species on this planet."

Edward Ellsworth
The Dolphin Network

Alpo Petfoods plans to include this "Dolphin Friendly" logo on the labels of cans of its tuna-based cat food by November 1990.

This "Dolphin Safe" logo will soon be appearing on cat-food products, including 9-Lives, Amore and Kozy Kitten brands, from Heinz Pet Products.

The survival of several Dolphin species(the eastern spinner and the spotted dolphin) are being threatened by the tuna industry. The National Marine Fisheries Service estimates that more than six million dolphins have been killed in the region from Southern Calif to Chile. It is in this region that dolphins and yellowfin tuna travel together. The tuna industries (both the United States and Foreign) locate the tuna by locating the dolphins and then encircling them both with mile long nets intended for the tuna but resulting in the drowning and cruel deaths of hundreds of thousand of dolphins per year. The figures do not count the young dolphins separated from their parents and the disoriented dolphins that escape but later die.

According to the Humane Society of the United States, this method of fishing for tuna devastates the dolphin population in many less obvious ways. Pregnant females and females with newborns are easy to catch because they are slower then the rest. It interesting to note that due to the numbers of fishing vessels in a certain area , many dolphins are attacked by this netting method only to escape and be attacked by another fishing vessel. This happens several times in one day! This constant harrassment and stress can only be detrimental to the future population.

Another interesting note is that only 10% of the worlds tuna is caught using the purse seine netting method described earlier on in this article. **Despite laws written to protect and control the dolphins this slaughter continues.** It is estimated that 800,000 dolphins have died in U.S.waters alone since the 1972 Marine Mammal Protection Act.

Due to recent consumer pressure some companies have become Dolphin-Free Tuna fish sold for human consumption.. This also effects dog and cat food. Several companies have begun to sell dolphin Free tuna for dog and cat food. Such companies as the H.J.Heinz Company, owner of StarKist Seafood Company, the largest American canned-tuna producer, and Heinz Pet Products and Alpo Petfoods Inc., the market leaders in cat food have announced that they would no longer buy or sell tuna caught with Dolphins (using purse seine netting methods). The company plans to label its tuna products including its pet foods, "Dolphin-Free".

Remember buy only Dolphin-Free tuna products!

Chapter 8

ORGANIC COCKROACH CONTROL

About CockRoaches.

Second only to ants in instilling hate and fear in all living things, cockroaches have been called " The Rats " of the insect world. CockRoaches belong to the order Blattoidea (family Blattidae), and can live up to a year. Females lay several egg cases containing 30-50 eggs ea during her life time. Some varieties carry their egg cases with them and place them for hatching later. Cockroaches utilize their antennae to pick up chemical signals from the air. This helps them in keeping away from any synthetic chemical used against them. What this means is that cockroaches will detect pesticides and avoid the area. They also sample food stuffs before eating to detect pesticides used against them. This is why using chemicals against them is fruitless. Cockroaches have as good a memory as ants. Like the ant, they too have developed immunity to many pesticides.

Since the first cave people, roaches have shared their home with humans. Cockroaches will live in almost any environment. They can be found on board ship, near food areas, bathrooms. Their eggs are good travelers, hitch hiking on board containers en route. Air travel doesn't bother them. They will thrive any where the moisture, temperature and amount of food available falls within tolerable boundaries.

A cockroach is capable of traveling many miles in one day. They can move through the cracks in walls. Where they go, their eggs will follow. A cockroach can live without food for three months and without water for over thirty days. Their egg cases can survive for years under the right conditions.

There are over 4,000 species of cockroaches known, 57 are found in the United States. Cockroaches have existed since the Paleozic era which is about 400 million years ago. Only 6 species in the USA are considered bad household pests.

On the CockRoach Trail

Identify areas in which Cockroaches are often seen. Keep notes. Take a daily count. Infestation occurs when there are too many to count. Determine tolerance levels. Is one Cockroach ok? Or is one too many? Keep count of population. This will also help you to determine if what you are doing is working or not. Set a max number you'll tolerate. If numbers reach max level then use the various techniques listed below.

Various Organic Cockroach Treatments.

Step 1

Using a Natural Cockroach Dust

20 Mule Team Borax

This is not a soap but a additive(laundry boaster) made from borax. It is a great natural product to use for the laundry but this product can be used with great success in controlling roaches. Simply sprinkle the borax powder in places where roaches are hiding(attic, insides walls, drawers, cabinets,car port, etc. Places where your kids , dogs, cats, etc can't get at. The main caution is to be careful where you put it so that it is not injested by anything other then the roaches. Use lightly.

A Safe Dust Formula

1 cup dried bay leaves(powdered),
1 cup peppermint leaves(powdered)
1 tablespoon of garlic powder,
1 tablespoon DE(garden grade)
1 tablespoon of cayenne pepper or chili pepper,
1 tablespoon powdered pyrethrium
1 tablespoon salt

Blend well together. Use a mortar to grind together into a fine powder. Sprinkle in locations cockroaches are seen. Spread a very thin layer. Follow along walls, in cracks, under structures.

NATURAL PEST CONTROL

Be careful using cayenne pepper as it will make you sneeze. Use is optional(works best along outside or unused areas). Use as much as you can handle. African Cayenne pepper is 940 BTU's and the best to use. Any type of pepper will do. DE should be used lightly inside. Works best if allowed to stay for 24 hrs. or more. In between walls, under houses, use in the attic, and in the cellar, are good places to apply. Put in closet corners.   Use outside along side of house and dust in places where they hide(under logs, wood, etc.). This mixture can be made into a paste by adding a little bit of water,and a tablespoon of butter, stirring until a paste like mixture is made. As a paste, you can place into roach cafes and place in the areas they are seen(but hiden). They will eat this and die.

Step 2

Using Soaps

Dr. Bronners Peppermint Soap is a safe natural soap. In use for over 30 years as a bath soap. Very interesting label, fun to read! Dr. Bronners makes several different types, peppermint, lavender, eucalyptus and others. The scent of these soaps makes them very effective against any insect.

The **Peppermint soap** is a very effective tool against roaches. It is an environmentaly safe soap to use. Spraying directly on roaches will kill them. Experiment on strength to use. Try 1 capful per quart. Use in kitchen and where roaches are seen. Spray along their paths to repell.

Shaklees Basic H will also kill cockroaches. They can not develop an immunity to this. Many people prefer this because it has no scent. A concentrate, so use a few drops per quart. Sprayed directly on roaches.

Citrus Soap is a safe natural soap to kill cockroaches with. Experiment with strength needed to kill them. Try a capful first.

Herbal Soaps can be a very important tool in your fight against the cockroach. Read the label first then try it out.

Let me know what happens!

Insecticidal soaps are made from fatty acids and are excellent for soft bodied insects. Safer Insecticidal soaps can be blended with any of the above soaps to increase effectiveness.

Tabasco Soap?

Tabasco sauce and soap works great as a liquid repellent sprayed outside. Simply add 2 tablespoons Tabasco sauce into a quart spray bottle filled with water and a tablespoon of a natural soap(such as Dr. Bronners Peppermint soap). Spray along where the walls of your house meet the ground. Spray also in areas outside where they hang out. This will keep them away. This mixture can also be sprayed directly on them to kill them.

Using Oils(extracts)

Coconut Oil will coat the cockroaches and will suffocate them. Add a dash soap to this and increase it's effectiveness. Experiment for proper strength.

Citronella is an incredible tool in cockroach control. Learn to use this oil against cockroaches. Using a small amount in water will repel them.

Garlic oil can be used against cockroaches with excellent results, (If you don't mind the garlic scent). Add a few drops to soap and water. Spray directly on cockroaches.

Melaleuca oil is a very strong oil to use against cockroaches. Use a few drops with a dash of soap and water. Test for strength.

Peppermint oil is another strong oil. This will kill cockroaches on contact! Test for strength and scent tolerance.

Step 3

Roach Traps

Sticky box traps use a sticky substance inside a open end box.They are called Roach Motels(tm). They can be purchased at most garden centers. Pop bottles make excellent traps as the fermentation of the soda inside attracts roaches and the pop bottle's shape keeps them from getting out, trapping them until they finally drown inside. Wine bottles, beer bottles and most long neck bottles will make good roach traps also. Always leave a little bit of the wine, beer,or soda in the bottle to help attract the roaches(or you can make your own Cockroach Brew, see this chapter).

NATURAL PEST CONTROL

You can add a little bit of soap to the left over soda,etc., in your soda bottle. Place the bottle in an area which they are seen, lean the bottle against something so that the roaches will have a way to climb in. Throw away the contents next day. Wide mouthed jars coated with petroleum jelly on the top 1 inch of the jar make excellent roach traps as well. Put Cockroach Brew into jar. Any thing like apple cider, wine, beer, sugar water, butter or dog food will attract them in and they can't get out! Roach Bait stations are made with boric acid and a bait (attracts silverfish too). A good roach trap is The Roach Eraser made by It Works Inc. Available at Peaceful Valley, Gardens Alive and others.

Making Your own roach sticky trap

Place a piece of bread onto a board which has been spread a layer of petroleum jelly or tangelfoot or you can buy fly paper and use that. Rat trap sticky paper mats work well also.

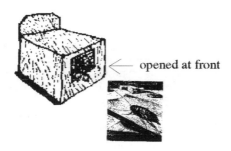

The COCKROACH INN

This unit is similar to the Ant Cafe or the Snail Inn. It is made from the same bird house(available at most pet stores) except it has a wider opening. Inside place a plastic cup. Screw down the lip. Inside the cup you can add the roach brew mixture. This unit keeps pets and kids out while allowing the cockroaches in.

← opened at front

Step 4

What do you feed them?

The Cockroach Brew

1/2 cup apple cider
1/2 cup cheap wine or beer or vinegar, apple cidar
1 tablespoon of Brewers Yeast
1 tablespoonful of Butter
1 tablespoonful Shaklee's Basic H
1/4 tablespoonful of Boric acid * optional if children, animals around.

Mix every thing together into 8 oz cup. Place into the Roach Inn. Screw down the lid. Keep children and pets out! Cockroaches will enter Roach Inn and go into the cup where the mixture is. Most die inside, others will die later. Change mixture and clean cup regularly. Keep count. Beer or wine, vinegar in a bowl, when used alone with brewers yeast really works well in attracting and trapping them. Try soda also. Keeping count will tell you how well the trap is working and how much longer to continue using. This trap will attract the young roaches also.

Epsom Salts (Magnesium Sulfate) any one ?

Fill bowl one-half full with Epsom Salts. Place bowl in area cockroaches are seen. Allow them access to the Epsom Salts by placing strips of wood against the bowl to allow the cockroaches to climb up. They will eat it. This will kill them within weeks. The magnesium will upset biologic systems in cockroaches and prevent them from feeding and they die.

Homemade Containers

You can use wide mouth bottles, bowls, jars, whatever is available. I do not recommend this if you have children. Make sure it's child proof.

* Note on Boric Acid....

If you have children then I suggest you use do not use the boric acid in any thing other then a safe place like in the Roach Inns. When screwed down children or pets will not be able to enter. Boric acid should not be ingested. Avoid cuts. Wash hands after use.

Boric acid powder is an excellent, safe (do not ingest! keep away from cuts, wear gloves) control of roaches inside the home if used correctly. 20 Mule Team Borax soap is such a product. Sprinkle into hiding places and along trails. Cockroaches have not been able to develop immunity to this substance, which they are unable to detect. They walk into it and ingest

NATURAL PEST CONTROL

through cleaning themselves. Boric acid takes several weeks to be effective. Keep children and pets away! Boric Acid is best used mixed into a mixture that is eaten by the cockroaches (see CockRoach Inns). Avoid breathing. I suggest that you try the mixture(s) without the boric acid first and see how well it works then use the boric acid if needed.

Drax works well on roaches as well as ants. Try Drax Peanut Butter for protein. Poisons have no effect on their egg cases and must be present when they hatch to be effective. Boric acid can be added to water to mop the floor with, added to water and sprayed around walls, etc., when dried will leave a thin film for the cockroaches to run in to(no cats).

Step 5

Vacuuming

Regular vacuuming of rugs, under furniture etc will pick up egg cases. Dust with pyrethrum / DE mixture 1/2 hr before. Dispose of vacuum bags by placing inside plastic bag and tieing shut.

Step 6

Getting a Tune Up

Our Daily habits must be changed to create less favorable conditions for cockroaches to survive in. Think about what you are doing and how it is promoting cockroach lifestyles.

Step 7

Storing Food

It is important not to provide food sources for them. there are many ways to store food. Using glass jars, canned goods. A seal a meal is a handy tool. Check your food containers. Throw away any food stuff with cockroach cases. Throw away into sealed plastic baggies.

Roaches can transmit salmonella, boils, typhus, dysentery to name a few. Throw away any food contaminated. Never leave food open over night. Remove stacks of newspaper,magazines and grocery bags. Throw away old books of no value. Look for good hiding places and remove them.

Step 8

Waste Management

Throw kitchen wastes into composter. Use large sealable trash cans. Spray Dr. Bronners Peppermint soap around trash cans. Spray on any roaches you see.

Step 9

A Little Handy Work

Keeping cockroaches out can be as simple as using rubber caulking around any entrances. Check around windows, doors, crevices, cracks, air vents, etc.. Check under house for possible entrances into house. Follow pipes into house. Check around connections, etc. Look for possible water sources.

Step 10

Modifying the Environment

The type of house we live in determines conditions which can lead to cockroach infestations. If you happen to live in a housing situation that promotes roach infestations, the first choice is to move. If this is not available then attempt to keep your immediate environment as clean and as healthy as possible to live in.

Keep your home clean and let lots of light in. Another way to modify the environment is to use incense. Try varieties that will keep them away.

Step 11

Cleaning Up

This is very important.If your environment feels good to you and encourages you to do better, then you will do better, and you will be able to control the cockroaches. Cockroaches have a well developed sense of smell. They will find any food that you leave for them that is out of reach of the ants. Cleaning Up is very important. To control them use Dr. Bronners Peppermint Soap when cleaning up. The scent will help to confuse them.

Growth Regulators

Gencor prevents roaches from reproducing. available from Gardens Alive!

Predators

The Gecko is a natural predator of the cockroach.

NATURAL PEST CONTROL

Many people are keeping them instead of cats. Chickens love cockroaches too so allow chickens to run outside if possible. There are many natural predators of the roaches from birds, spider, lizards, wasps etc. Allways encourage them by providing a safe place for them to be in.

"Remember, give yourself time to control the roaches"

SEPTEMBER 24, 1881.]   PUNCH, OR THE LONDON CHARIVARI.

"SO NEAR AND YET SO FAR!"

BIRDS ARE VERY WILD. TOMKINS, WHO IS NOT IN GOOD CONDITION, HAS FOLLOWED THIS COVEY OVER FIVE FIELDS. SOMEHOW, HE SAYS, HE CAN'T GET NEAR 'EM. THEY SEEM TO HEAR HIM COMING!

NATURAL PEST CONTROL

Chapter 9

Organic Fly Control

About the Fly

Flys have been around before mankind. There are many different groups of flies, from the common house fly to the hover fly. But when we think of flies we think of filth flies. Flies which are associated with human-generated garbage and animal manure. Many species of flies are predators or parasitoids of other insects. Their larvae eat decompsing animal bodies. Flies carry many different diseases which can effect humans as well.

There are over 85,000 species of flies.

Here are some well known varieties of Flies

Fruit Flies: Live in ripe fruits , garbage, and rotting vegetables. Proper Cleaning of the area is important here. Keep Garbage cans on tight and remove all rotting materials.

Black Cherry Fruit Fly:
prefers sour cherries.

The Cherry fruit Fly:
prefers cherry, pear, and plum.

The Currant Fruit Fly:
prefers currents and gooseberries.

The Oriental Fruit Fly and The Mediterranean Fruit Fly:

Have a wide range of tastes from many citrus, fruits such as peach, nectarine, plume, grapefruit, orange, apple, pear, coffee and many more fruits and vegetables.

The Mexican Fruit Fly:
A pest of citrus fruits, mangos, and many other fruits.

Carrot Rust Fly:

The Host plants are Carrot, celery, parsnip, and parsley. Maggots cause damage to root system. Soft-rot usually ollows. Maggots winter in soil.

The Common House Fly:
prefers garbage, manure rotting fruits, vegetables.

Syrphid Fly:

They hover like bees. Attracted to tree sap and fermenting fruit, are predaceous and feed on aphids, mealybugs and other insects.

Tachinid Fly:
beneficial. Parasitic on other insects.

More about the FLY

Habitats:

Primary food sources for filth flies is human waste such as garbage, and pet manures.

Growth Cycles:

Goes through four stages in life cycles

Egg, Larva, Pupa and the Adult. If you interrupt any part of the cycle, you have gained control.

Steps to Control

Step 1

Control fly population thru proper disposal of kitchen and other wastes, and other recyclable wastes, and also thru pick up & dispose of your pets feces. Keep flies out of the house by using screens, bead curtains. Keep windows and screens in order. Use Fly swatters. Use fly paper. There are many different types of stick-ems which you can hang up that will trap them. Reduce the sources of infestation and you will have a good hold on the fly populations.

Organic Fly Control

Step 2

Using Soap to Control Flies

Use soap to spray directly on flies and to spray on manure. Use either Shaklees Basic H or Dr. Bronners peppermint soap. Use 1/2 cup per gallon water. Mist area where flies are seen. Great for using inside the house. Just mist the area and allow the flys to fly thru the mist. This also wets their wings and allows you to swat them.

Using soap and Tabasco Sauce

Use 1/2 cup soap and 10 tablespoons Tabasco Sauce per gallon water. Spray directly on places where flys can lay their eggs. Not for use inside as the Tabasco will stain.

Step 3

Using Natural Dusts

DE(garden grade) can be dusted on the manure to control flys. Also DE can be fed to your animals, horses. How to use DE is explained in other parts of this book. Use 1% of food weight. For dogs and cats feed 1 tablespoon per feeding. For horses feed 1/2 cup per week added to their meals. The DE when feed, will prevent flys from laying eggs in their manure. The DE will also help to deworm as well as provide many trace minerals. You can also dust the manure with the DE to control flys that way.

Pyrethrum Dust

Dust around areas where flies are hanging out. Can be made into a spray by first making into a paste by slowly adding water to the pyrethrum then adding that to water(see appendx for more info). There are many liquid pyrethrum products on the market,use only liquid pyrethrum that does not contain any additives. An excellent mixture is DE and Pyrethrium. Mix 50%

Step 4

Biological Control

Use parasitoids of flys such as Spalangia endius. muscidiflurax zoraptor, Pachycrepoideus vindemiae, Tachinaepheus zaelandicus. Use beetles, mites or soldier bug.When using parasitoids do not use any sprays as it would hurt them also. Do not use DE either. Best to leave them alone to do their job.

Organic Controls of the Medfly

Medfly eradication efforts are proving to be a great waste of energy and money. Concentrating on the effect instead of the cause is wasted effort. Medfly spraying of Malathion or any other chemical is wasted effort and will not effect the real cause of the Medfly spreading in southern Calif. Medfly infestations are spreading through-out the world!

Even this is not the real cause of the Medfly problem. The real cause is chemicals and the imbalances they have wreaked upon our planets eco-system.

In order to regain balance we must stop relying on chemicals to control our pests but instead rely more on maintaining a balanced eco-system, diversity rich with bacteria and enzymes. We must rely on sustainable organic methods of food production. We cannot depend on chemical control and the loop it causes forever. This system will break down and cause chaos and disorder,

Re-Education of our farmers, retraining of professionals, reteaching the teachers, is necessary. Our children must be taught the best way to grow and enjoy nature. Organics and the art of living without chemicals must be taught to all of our children and to their children and made into a way of life.

This is all well and good but what about now, what do we do now you say?

Stop spraying any kind of chemicals.

NATURAL PEST CONTROL

Start holding workshops to reach people on the various methods of natural Medfly control.

Start workshops for farmers who can be effected by the Medfly. Show them natural methods of Medfly control.

Encourage organic pest control methods with professionals by allowing for a natural/organic license for those wishing to use only organic methods of pest control.

Encourage more research on developing natural predators.

The best defense is a good offense. Provide for your property plenty of good rich compost, use only natural fertilizers and depend on natural predators to keep things under control.

There are many natural predators of the Medfly. Many insects attack the fly and/or it's eggs. Birds, toads, and other predators can keep fly populations down. Reduced use of chemicals increases functional diversity of life. It is this diversity that will maintain the balance.

DE can be used to control the Medfly. DE can be added to water and sprayed on the fruit(s) affected by the Medfly. DE can also be sprayed on the ground around the fruit trees affected. Should be done in the spring.

The following formula can be used to spray on most fruits and vegetables to control various flys

1 cup DE
20 drops natural soap such as Dr. Bronners Peppermint soap, Insecticidal soap, Citrus soap, Coconut Oil soap, Shaklees Basic H (any natural soap will do, experiment for best results).
5 drops Garlic oil,
10 tablespoon tabasco sauce.

Mix well in gallon container. Add water to fill. Stir well. Allow to settle and strain into sprayer. Spray on fruit and allow to dry. Can be sprayed directly on Medflies and other flies. Will also control flys and other insects so be careful where you spray.

Another Formula

Take 1 lb compost and 1/2 cup DE, place into panty hose and place that into a gallon of water in a glass container and allow to sit for three days in the sun. Add 20 drops soap. Add to sprayer and spray fruits, and other areas.

SOME ATTRACTANTS & TRAPS AVAILABLE IN CONTROLLING FLIES

SAFER(TM) HOUSEFLY TRAP

This product uses a time released BioLure(tm) which attracts the flies into the trap.Comes as a kit and contains the following:1 Safer(tm) HouseFly Trap with top, cup, lure and bag attachment ring, two collection bags and one BioLure(R) in plastic bag. This trap is easy to assemble and instructions are easy to follow. Place at least 4 to 6 ft above the ground. Hang from eaves, roof, ceilings of barns and kennels or hung from trees or shrubs. Place in shade during summer and in direct sunlight in cooler weather. Place near sources of infestation such as kennels, garbage, horse manure areas, etc. Trap should be serviced every two weeks by replacing lure or bag if damaged. You can purchase additional lures from the company. The additional plastic bags however can be bought to fit the trap.
Safer Housefly Trap is available from
Safer Inc.(see resource directory).

Comments:

After trying this product for several months, I am impressed with several things: The unit itself looks great and is easy enough to put together. The lure which it comes with works very well as long as it is not allowed to dry. This is the main problem with this unit. The conditions under which this unit is operating will determine how often you will have to check the lure for water. Also , in order to increase its effectiveness, I added a small amount of fish emulsion into the lure's small container, and a small amount of water to that with a drop or two of soap. The fish emulsion really attracts the flys but can cause quite a smell so don't place inside the house. Also a small piece of fish can be used here.

NATURAL PEST CONTROL

The Fly Terminator

This product has been around for a long time. Effectively attracts and kills flies without poison scatter baits or electrocutes. Uses attractants. This is a large bottle looking product which holds water. An attractant such as a piece of fish or fish emulsion can be used instead. The instructions say that no meat or fish is neccessary because of the attractant they sell with it, however I found it preferable to use fish heads, etc., and/or to add a small amount of fish emulsion to the water and to change it every month or so as needed.

The unit is placed or hung near where there is a fly problem. I have found this system to really work in catching attracting and killing flies.

Peaceful Valley Fly Trap

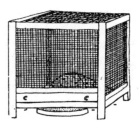

A trap for catching large numbers of manure breeding flies. Trap will hold up to 25,000 flies. Non toxic yeast and ammonium carbonate bait attracts the flies.

Fly Sniper

This product uses a sex pheromone to attract the flies.

Rescue Fly Trap

This is a handy disposable model. Just add water and hang. Throw away when full.

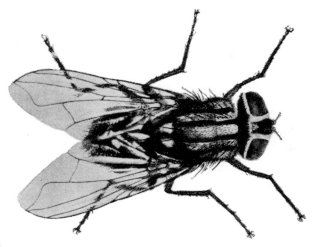

Page 64

CHAPTER 10
NATURAL TERMITE CONTROL

Natural Termite Control

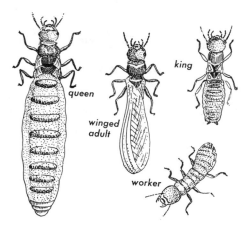

While there are over several hundred varieties of termites in the world, only four live in the United States, subterranean, drywood, dampwood and powderpost. The subterranean being the one that causes 95% of the damage. Subterranean require contact with a moist soil. This limits their attack on wet and dry woods within reach of the soil. They make tunnel like tubes to connect the soil and wood together. These tunnels are meant to protect them when they travel and are usually found free standing in the crawl space nelow your home.

Drywood termites don't require the moisture that subterraneans do. They can attack the structure far away from the soil. They create small galleries with small entrances pluged with partially chewed wood and a cement like secretion. Drywood termites leave piles of sawdust like pellets.

Termite Prevention

The conditions that invite wood damaging termites can be discovered and corrected long before the problem gets out of hand. Therefore it is important for the home owner to learn the proper measures that can be taken to prevent termites from entering your home.

Step 1

Inspection

Poor Home Maintenance

Repair loose or cracked siding or stucco, peeling paint and gaps around windows and doors which allow moisture into the wood which provides an attractive environment for them to live in. These exteriors should be properly repaired or replaced. Peeling paint should be removed and replaced. Adding boric acid to the paint will increase its effectiveness.

Low foundation walls which allow close to the earth to wood contact are the main areas to look out for as these are prime conditions for the termites to enter the house. The foundation should be raised or the earth lower at least 6" preferably 12". Repair all cracks in the foundation that provide access to the wood inside. Look out for the gaps at delaminating brick veneers which provide easy access to the wood below. Repair or replace. Must be water tight.

Poor ventilation in crawl spaces will encourage moisture and dampness will occurr which is again great conditions for the termites to live in. Insure crawl space vents for proper ventilation of all areas. The vents can also be dusted to dust the area with. I will explain this later on dusting. Check for plant growth directly outside vents to insure that it's not inhibiting proper ventilation. Remove or trim any brush etc as needed. Always avoid plantings that are closer then 2 feet from the house. A good rule of thumb for for crawl space ventilation is about two square feet of opening for every 25 lineal feet of wall.

A MISSOURI BARN.

NATURAL PEST CONTROL

Firewood Storing

Avoid storing your fire wood or wood scrapes next to the house as this could support termites and allow them entrance into the house. Also remove any old tree stumps which you may have around as this may become host for the termites.

Plant Container Control

Avoid plant boxes attached to the house unless it's off the ground. Control the watering to avoid constantly wet wood. Soil nematodes can be added to the planter boxes to descourage termites. See Using nematodes page.

ANCIENT HOUSES IN HENLEY STREET 1820.

Structure control

Porches, decks, and other wooden structures should not be in direct contact with the soil. Concret footings should be used. I suggest that you use builders sand around the base of any wooden structures at least 2 feet out and two feet deep. Sub termites will not pass thru this barrier. I suggest that you mix DE with the builders sand as you apply it. See barriers page.

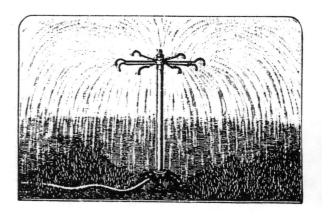

Water Control

Leaking pipes or water faucets will keep the wood moist causing damge to the wood and providing access to the termites. Look for leaks. Check the water meter after turing off all water use. If still moving then you have a leak. Leak control is very important in controlling termites.

Control your watering via drip or soaker systems. Over watered landscape is a major part in attracting termites. Provide proper compost and mulch to insure healthy balanced soil bacteria. Use Nematodes to control. See nematodes page.

NATURAL PEST CONTROL

Ant or Termite?

Can you tell the difference between ants and termites?

	Ant	Termite
Antenna	Elbowed	not Elbowed
Body	middle part of body very narrow	middle part of body not narrow
Wings	wings not alike in shape size and pattern	wings similar in shape, size or pattern
Veins	few veins	many small veins

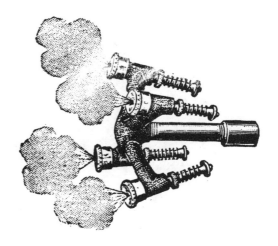

Other Considerations

Pay special attention the the bathroom , kitchen and laudry rooms as these are the places most of your water is used and a great number of problems arise. Check for leaks and repair or replace as needed.

Step 2

Once you have found signs of wood damage and you have determined that its not just old decaying wood then you have to decide if it's termites or carpenter ants that you have. There are also termite inspecting Dogs called TADD Dogs.

If you have determined that its carpenter ants then go to the Dances with Ants chapter and treat likee you would the ants, the only difference would be that you place the ant cafes inside the house, in the attic, crawl spaces, any place where they are seen, hiden from view.

Methods of Control

Do it yourself

Aside from inspecting your home yourself for obvious signs of termite damage, there are many things you can do yourself:

Step 3

Dusting

Here are some natural dusts you can use around the house to control termites. Dust in places such as attic, crawl spaces, around outside of house where wood meets the ground or other moisture.

DE

DE is one of the safest methods of providing termite control. See DE page in Appendix. De can also be mixed in with your builders sand for added protection. DE can also be painted on the wood. I suggest 1 part DE to 1/4 part Boric acid, added to water until a thick paint like texture, then painted on the wood.

Pyrethrium

Pyrethrium can be added to the DE and painted on for additional protection. Best used as a dust for immediate contact kill. Mixed with equal parts DE for an exellent dust.

NATURAL PEST CONTROL

Step 7

Barriers

Construction Sand and DE mixed together and layed down around the house 1 foot deep and 1 foot out from the house will prevent termites from moving into your house.

Step 4

Spraying

Spraying of attics and other crawl spaces with a liuid DE/Boric acid mixture is highly recommended at least once per year in the early spring.

Step 5

Surgery

Removal of their colony and tubes is easily done upon detecteion. Remove infected wood and replace with newly treated wood(you treat yourself with the Boric acid/DE paint). Look for signs of their tubes and removal these. You can also dust the areas.

Step 8

Nematodes

Termite eating nematodes are available thru any of the mail order houses such as Arbico and Peaceful Valley Farm Supply. They are easy to apply(using a garden hose and attachment unit) and are safe as well. Should be used on a yearly bases to insure effectiveness. Works only on subterrean termites.

Step 6

Boric Acid

Boric acid can be painted on to the wood as a form of protection against termites eating into it. You can also use 20 mule team borax to dust into their tunnels. Avoid breathing.

Ant Friends

Termites and most ants are enemies and will fight each other. Carpenter ants are often mistakened for termites(see below) and such be avoided but most other ants can be allowed to live outside(provided they don't come in(see ants section for proper control).

NATURAL PEST CONTROL

Step 9

For Hire

If you have done the above and still have termites then I would suggest this section.

Commercial methods available for termite control

Heat

In which propane heat in a tarp wrapped house is used to raise the temperture of the wood to 120 degrees(room air to 187 degrees) This method is called Thermal Pest Eradication..

Effective rate 95%

limitations for best results should be done every 5 years(more often in tropical climates). Best combined with another type of treatment such as Liquid nitrogen for spot treatments.

Liquid Nitrogen

292 degrees below zero, is sprayed into walls. pinholes

effective rate 100%

limitations cannot reach all parts of the hous, best combined with another type of treatment such as electricity. The Tallon company uses a system called The Blizzard.

Electricity

using an electrogun, electric shock(90 watts-90,000 volts) is applied to the walls and other areas of the house thru tiny pinholes which are drilled and repaired after use. Pentrates up to 1 inch into the wood. Nails can be used to extend its reach. Look for a company called ElectroGun.

effective rate 80%

limitations: Cannot reach all parts of the house best combined with either liquid nitrogen or heat or micro wave. Used as a spot treatment.

Microwave

500 watts is applied to the house thru a funneled device the size of a toaster oven setup on a tripod close to the wall.

effective rate 90%

limitations cannot reach all parts of the house best combined with another system such as heat or electricity.

Chapter 11
ORGANIC LAWN CARE

The dangers to our health from the chemicals being used to keep our lawns green and weed free are increasingly becoming more apparent. Strangely named chemicals such as pendimethalin, benomyl and 2-methylcydohexy are widely used to combat weeds.

There is an ever growing list of chemicals used to maintain a "healthy lawn" along with the use of chemical fertilizers, herbicides, pesticides and who knows what else, causing concern amongst the public. These chemicals not only have dire effects on the environment but on we 'humans' as well. We can no longer trust the judgement of the EPA.

"Being registered with the EPA by no means counts for a seal of approval or a seal of safety", says the National Coalition Against the Misuse of Pesticides.

Therefore, it is important to begin to take it upon ourselves to control the chemicals we use in our everyday lives, such as the chemicals we use on our lawns. This chapter is about how to grow a deep green healthy lawn without the use of man made fertilizers and pesticides.

We will discuss Preventative Cultural Practices which is the foundation a healthy lush green lawn is based upon. Understanding the components of the lawn's ecosystem and providing a more suitable environment for the types of grasses and micro-organisms which grow best there while decreasing the conditions for pests and harmful bacterias, is the "ideal" and more natural, less toxic method of providing yourself with a beautiful lawn.

Preventative Cultural Practices

PCP includes the proper selection of the types of grasses for your area and the lawns use (are you going to play on it, walk on it a lot, or just enjoy its beauty)? Having a healthy lawn without harsh chemicals is really very easy to do. Lets go over the basic steps together.

Step 1

Understanding The Soil

When you understand the type of soil in which your lawn is to grow or is already growing, then you can begin to create a soil structure which is better suited for proper plant growth. Soil supplies air, water, nutrients, and physical support to plants. Soil also needs organic matter to keep it alive.

Each type of soil needs a different amendment. Sand, clay, and silt, in varying amounts, determine the texture of the soil. Compost will help to bind soil particles and hold water to the soil. Water is also important for productivity. Too much or too little water hinders proper plant growth. In addition, the presence of earthworms aids in maintaining good soil structure.

Use only organic materials. Increase organic matter. Use manure as a top dressing. Use organic fertilizers. Lawns love compost. Lawns love rock dust. Blend the two together for best results.

"There are two ways to sterilize the soil-by Heat and Chemically"

Chemical fertilizers kill off the beneficial soil bacteria, as well as killing off earthworms and therefore should not be used, should be avoided at all cost! In this environment, the lawn and its bio-system will be operating under stress. Stress is the most important factor in pest control.

Whenever the lawn is under stress, several things have occurred: 1: Proper Food is not being made available to the lawn. 2: The Humus levels have dropped below optimum. 3: Water effects stress. Too much or too little water will cause stress in the lawn(chlorinated water is bad for the soil). 4: High nitrogen fertilizers kill the soil and cause a great deal of stress. Avoid using Urea at all costs.

NATURAL PEST CONTROL

Understanding The Environment

More than any other element of climate, temperature will determine which type of lawn will grow where. Available sunshine and water (either in the form of rain or humidity) round things out.

An example of this is: if irrigation is unavailable, only certain desert species of lawns can be grown. It is a good idea to consult a gardening book which has a map broken into climatic zones. This will help to simplify the process of deciding which grasses you should consider.

Step 2

Soil Amendments

There are many forms of organic matter. Untreated sawdust, aged horse manure, and aged wood are common amendments and can be found anywhere. Aged-wood soil conditioner, cocoa bean, mushroom compost, rice hulls or apple pomace are usually available regionally. Organic waste from your kitchen can be used to compost your soil as well as additions of rock dust.

Other amendments are: leaf mold, pine bark and bark chips, straw, sand, coffee grounds, grass clippings, shredded cardboard, bat guano, eggshells, grapefruit skins, potato skins, and wood chips and ashes.

Organic matter should be added to your soil before you start a new lawn. Should be done when the weather is fairly settled, so the ground will be prime for planting and the new grass will have only mild competition from weeds. The Best Organic matter to add is in the form of Compost. See chapter on compost.

Organic Fertilizers

Soil, in order to be alive, must have high organic matter, drainage and good structure. Natural fertilizers add to the soil, improving its fertility,maintaining and contributing to the improvement of these necessary elements. Organic matter and rock powders form the basis of organic fertilizers, and benefit the soil as well as the plant. As a rule, chemical fertilizers are not a complete plant food. Organic material contains nutrients which provide the microorganisms in the soil with the materials they need to be active.

Organic Fertilizer mixes usually contain composted animal manure, plant residues, seaweed and fish products, and minerals (bone and blood meal, cottonseed meal, granite dust, phosphate rock and greensand). Organic Fertilizers should be used in combination with compost in order to develop rich, humousy soil.

Compost: Composting is practiced today just as it was hundreds or thousands of years ago. The recirculation of dead matter into life is a part of nature's program of soil rejuvenation.

A gardener's compost heap is a process that is going on eternally in nature. When we cut the grass and remove it we are cutting off the cycle. Therefore compost, when added back to the lawn, re-establishes the cycle and returns nutrients and bacteria to the soil. The gardener takes a tip from nature and uses this simple method to build the soil's fertility.

"*Compost is the Finest Natural Fertilizer there is*"

Horse Manure

Among the manure of other farm animals, horse manure is the most valuable. It is richer in nitrogen than either cow or hog manure, and ferments much more quickly, therefore being referred to as "hot manure".

Manure provides organic matter and trace minerals to the soil. Your lawn will benefit by having a thin layer of aged (at least 6 months old) horse manure spread over it (using a manure spreader), and then watered well. This should be done twice per year or even seasonally if you can do it.

This practice will also benefit the microbes as well as the beneficial insects. But the best top dressings are made from well made compost because it provides a more complete and varied food source as well as varied bacterial sources.Run through shredder for best results.

Page 71

NATURAL PEST CONTROL

Step 3

Watering Practices

Deep, regular waterings are essential to develop a lush, verdant lawn. Lawn sprinklers are a definite improvement over hand watering. They can be set in place to water as long as necessary, and are especially effective if placed on a timer.

The amount of water used depends on how deep the roots are of the grass you are growing. If there is a drought, it is better, rather than sprinkling lightly, to water twice a week, soaking the soil to a minimum of 4 inches, or not at all, for shallow watering causes roots to spread out near the top where they will be baked by the sun, becoming unable to withstand drought.

A lawn which is watered thoroughly at regular intervals, and whose soil has plenty of organic humus (organic matter converts into humus which makes the nutrient elements in the soil available to the grass), will withstand drought, and remain sound throughout the hot summer months. Different types of soil need different amounts of water. Light soils need more water because they drain so rapidly. Clay soils retain water, so they need less.

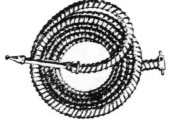

Step 4

PEST IDENTIFICATION AND TOLERANCE LEVELS

Not all pests are bad for the lawn in low levels for they perform a desired task in maintaining a balanced ecosystem. This applies to weeds as well as insects. 5 to 10% weed growth is allowable and often not noticed. This also depends on whether you are offended by a certain type of weed and what weeds don't bother you.

This will make a difference to you as to the over all appearance of the lawn. Decide what levels you will tolerate and operate within them. As for pests, develop a monitoring system which will allow you to determine what steps are needed to keep these pests below your tolerance level. Learn to identify the good guys from the bad guys. This is important! There are lots of good bug ID books out there.

Step 5

MONITORING SYSTEMS

Monitoring is when you regularly inspect the lawn for signs of harmful and beneficial lawn pests. Before you cut the lawn, you should inspect it for signs of stress. Here a garden notebook becomes very handy. Write the date and location and what signs you have observed and also write down what you did to correct the problem.

Step 6

DEVELOPING A REGULAR MAINTENANCE PROGRAM

It is important that you decide what work must be done on what basis, and do it. Watering, fertilization, pest control, and maintenance should all be done on a regular basis, and notes should be kept on the whole process.

Here are some maintenance steps which you should follow
Soil Test... once per year.
Organic Fertilization... twice per year.
Top Dressing of lawn with compost... every season.
Irrigation... establish a regular program of deep waterings on the same days at least two or three times per week. Watering early in the morning is usually best.
Aeration depends on soil structure and should be done once per year.

NATURAL PEST CONTROL

Maintenance... Cutting the grass is a very important step. Cutting too short or too long depends upon the type of lawn and the type of lawn mower.

Mulching Mower

Using a mulching mower is one of the best ways to recycle your grass clippings and feed the lawn at the same time!

Step 7

Identifying Lawn Problems

Pests of the Lawn

Caterpillars/Grubs

Caterpillars/Grubs are found in the soil. Grubs cause damage to the lawn by leaving yellow or brown patches(roots of lawns are eatened).

Soap Drenching

You can check for grubs by placing a can that has be cut at both ends, place one end into the lawn then pour soapy water into in. In about 10-15 mins, see if any grubs have come to the surface.

These grubs can be dealt with the use of beneficial nemataodes. Just follow instructions for its application. In most cases its a simple matter of dissolving in water then spraying on areas. Compost applications and avoidance of high nitrogen fertilizers will help to control in future. Beneficial nematodes are highly recommended for this control. Try Arbico or Peaceful Valley.

Japanese Beetles love poor dead soil since their natural enemies are not present. The Grubs of these are a very serious threat to most lawns. A natural lawn will not be attacked by the beetles as an over chemicalized lawn would be. Benificial nematodes provide excellent control. Milky spore disease is highly effective against this and most beetles.

Chinch bugs are also serious pests of over chemicalized lawns. Use the soapy soil drench method as described above to see if any chinch bugs come to the surface. If you think your lawn has chinch bugs then you must first pay attention to proper caring for your lawn. Follow the basic steps mentioned in this chapter for organic lawn care. Top dressing of compost is a must.

A Natural Chinch Bug Spray

Mix together Shaklee's Basic H , Dr. Bronners Peppermint Soap, Citrus soap and Safers Insecticidal soap. Mix equal amounts of each. To a gallon sprayer add 1/4 cup and spray on the lawn. Best to spray in late afternoon to avoid any damage. Don't spray during a hot day. Water in well afterwards.

Moles in the lawn

See chapter on gopher control and follow same procedure.

Gophers in the lawn

See chapter on gopher control.

Ants in the lawn

see chapter on Ant control.

Additional information on controlling ants on your lawn:

Follow the procedures described in the ant chapter but also use the following on your lawn.

Steps to remove ant hills from your lawn

Step 1......to one gallon water add 1 cup Shaklees Basic H and pour into ant hills.
2.....mix equal amounts of DE and Pyrethrium dust. dust hills, turn over with shovel to un cover ant colnies.
3....as last resort pour boiling water into ant hills.

Mining Bees in the lawn

Their mounds can be removed with a rake. They do very little damage to the lawn.

Dogs peeing on the lawn

Hose with water asap. Top dress with compost.

NATURAL PEST CONTROL

Step 8

Lawn Damage

Drought Damage is caused due to improper water and composting of your lawn. Watering deeply less often is best. Areate the soil to allow water to enter.

Chemical spills can be cleaned up with a little citrus soap and water. Hose down well after wards.

Compaction is due to improper areation and watering. Top dress with lots of compost or old horse manure once per year.
Fertilizer burns can be helped with lots of watering to flush away chemicals. Top dress with compost to allow lawn to come back.

Pesticide burns should be treated the same as above.

Step 9

Diseases of the Lawn

Algae growth generally does very little damage to lawns. Main cause is over watering and high nitrogen fertilizers. Top dressing with compost will help conditions.

Fungi Diseases

Fungal Diseases are caused by dead soil and using high nitrogen fertilizers. Over chemical use on lawns eventually will kill off all the beneficial bacteria in the soil, leaving proper conditions for the bad guys to move in. Best type of fertilizer for the lawn is properly made compost, or other natural fertilizers such as Rock dust, Nature meal(Nitron), and many other natural lawn fertilizers available on the market today. Use Nitron A-35 as a bacterial activator (also try AgriGro, SuperSeaweed, Shure crop, etc). Wettable sulfur makes a good immediate fungal control.

" Disease can be avoided by proper fertilization and proper watering techniques"

Lichen growth can be reomoved by hand and raked clean. Top dress lawn with good rich compost and spray nutrition such as Superseaweed, Nitron A-35, Fish emulsion, etc.

Toadstools or mushrooms are not neccesarily to be viewed as bad for the lawn but instead it should be looked upon as an indication that the conditions of the lawn should be looked at more closely. If your lawn has toadstools then pick them and dispose. Top dress the lawn with good rich compost. Keep an eye on your watering. Water in early am is best.

Step 10

Natural Weed Control

Weeds will only grow in soil that is not rich in minerals. Weeds need minerally deficient soil to grow in, so keep your lawn well fed with minerals and natural bacteria. You will have very few weeds establishing themselves in your lawn.

Proper weed control is established thru:

Proper mowing of lawns

Always pull up weeds before you mow to avoid spreading the seeds. Keep a sharp blade.

Correct watering

Regular deep watering is better then often shallow watering. Watering in AM is better then watering in PM.

Correct Fertilization

Stop using chemical fertilizers. Stop the high nitrogen cycle. Organics will provide plenty of nitrogen etc as needed by the lawn for healthy weed free growth. Give lawns plenty of rock dust which provides minerals needed by healthy lawns. Provide your lawns with a top dressing of old horse manure or aged compost at least yearly. Chicken manure top dressed on your lawn 4 times per year will keep your lawn green and healthy. There are many natural fertilizers available on the market today.

Making your own Organic Lawn Fertilzer

5 lbs Alfalfa meal
5 lbs Rock dust
10 lbs compost (Fine grounded)
10 lbs chichen manure or LLama manure or rabbit
manure (fine grounded)
5 lbs coffee grounds

Mix well together. All items should be run thru a fine screen. Apply as needed.

Making your own liquid lawn food for spraying

Nitron A-35
Superseaweed
1/4 lb Rock dust
1/4 Alfalfa meal
1/4 Seaweed meal
Mix together and place into a panty hose tied into a ball. Allow to sit in filtered water or solarized water for an hour and spray onto lawns. Best time is early am. See making your own foliar spray chapter for more.

Hand weeding

Hand pulling of the weeds is an easy and natural way to control weeds. I suggest that you first spray the weeds with a mixture of Shaklees Basic H and Safers Insecticidal soap(50/50) the day before(see below). This makes pulling the weeds up easier to done. Weeds can also be sprayed with this mixture and left in the ground. They will decompose and provide food for the soil. Test the strength of the soap mixture. Make to strength as needed.

Using Natural Soaps

Weeds can be controlled by using natural soaps. Soap will burn them and they will die without having to remove them from the soil.

Shaklees Basic H/ Safer Soap

Mix these two soaps together at 50/50 ratio. Add 1 cup of this mixture into a gallon sprayer. Should work on most weeds but will not damage the lawn. Don't water for 24 hrs and don't spray in midday or on a hot day.

Getting Started

Step 1

The first thing you have to do is address the nutritional needs of your property as well as that of the plants. Proper nutrition is the KEY to natural pest control. Anytime you treat the cause, you will be controlling it's effects. Ask yourself "what am I doing to the soil?". Stop using chemical fertilizers. Stop using pesticides that are not made naturally. Avoid high nitrogen at all costs.

Yes, you will have to switch to Organics if you want to regularly control thrips, aphids, and other pests naturally. Organics and chemicals do not mix well. You may get away with good results for a few years longer then if you were completely chemical but sooner or later you will pay for it, in the form of one pest or another, or one disease or another, not just the plants and soil suffer; you will also suffer from exposure to the chemicals in the fertilizer or pesticide, etc.

Fertilizer exposure can be as dangerous as pesticide exposure. So decide now if it's organic that you want to do it. Remember, you can't be organic only when it suits you to do so. Organic demands it to be your way of life.

Take Good Care of The Soil

If you are a first time rose grower there are many things we must agree upon before starting. How much time do you plan to spend on your roses? Are you a part time rose lover or are you going to really go at it? There are many good books available on the subject of growing roses.

Take an organic rose growing class. Growing roses for show is a great deal of fun and also lots of work. You will also have to decide if you are growing organically or if you will be using chemicals.

Organic or Chemical?

Our health is the number one reason for growing Roses organically. Consider the many chemicals on the market today and the many health problems that arise from their continual use. Remember that children and older adults are at greater risk. Secondly, consider the environmental pollution associated with their production, transport and consumer use. You can have that perfect show rose grown organically and stay healthy to enjoy it. An added benefit is that roses organically grown are more disease resistant, & they last longer!

The Big Switch

Going from Chemical to Organic can be a traumatic experience to both the rose and the grower. Not only Plants but people go through chemical withdrawals. You must allow at least one year to complete the switch(see converting to organics). This allows both you and the plant time to adjust to the new organic regime. While it is best to start off organically, you may not be able to. Commercial rose growers are not organic and therefore when you first purchase your roses they have been grown chemically! Don't Panic! Given time your roses will love being grown organically.

Which Variety?

The varieties you choose are important. Environmental conditions determine what varieties will do well and what will not. Don't just plant any rose or you will be sorry! Take into consideration the following: Is it hot and dry? Is is very wet? Fog? Winds? Where are you located? Your state will have its own special environmental issues that you will have to deal with. What varieties are being grown in your location?

Location

Choose an airy location to allow for air circulation to reduce fungus development. A sunny location is a must if your going to have stress free roses. Early morning sun is best for roses.They like to look at sunrises! Give them as much sunlight as possible.

What Condition my Condition was in

The soils condition determines everything. If you have all of the other conditions correct and not the soil, you will not be able to grow them organically and chemicals will be needed. The key to growing anything organically is healthy soil. Living soil allows for a functional diversity of organisms. Functional diversity is how organisms cooperate with each other. This sharing of resources provides for an ecologically balanced system. The greater the balance the higher the energy level, the greater the nutritional levels, the less

NATURAL PEST CONTROL

the stress and the less the pest.

Know your soils PH

Different plants require a different PH for maximum benefit. Know your plants and their PH requirements. In soil where the PH is too high it will not be able to assimilate certain minerals resulting in a trace mineral deficiency which invites a pest or disease attack. Most animal manures have a high PH and you must use peat moss, forest shavings, etc., to counter balance this. Rock dust also has a high PH(8-9 PH) and therefore must be balanced with a slightly acid mulch/soil, compost. Roses love ph of 6.5

Step 2

Watering

Roses do not like over head watering. Always provide either a soaker hose or a drip system for them. Regular watering is important. Avoid over watering. Slow deep watering is the best. Use a 2 gallon per hour drip head, one on either side at apprx 6 inches out. A soaker hose will work fine for this purpose as long as it doesn't spray water onto the leaves. A soaker hose can be buried or mulched over to prevent this. Dig a well around the rose to hold in the water and the compost and mulch. Mulch should be 2-3 inches deep. Fertilizing Your Roses:

Avoid using high nitrogen fertilizers. Nitrogen is naturally provided in the organic system and is never lacking. In the organic system, nitrogen is easily available and only used when needed and in small amounts for longer lasting results. Some natural sources of nitrogen are animal manures such as horse, cattle, llama, rabbit, earthworm casting, chicken etc.; they also provide natural bacteria, enzymes, and trace minerals.

Step 3

Nutrition

Compost and Roses

Get good at making compost. Experience is the best teacher here. The better your compost, the better it will be for your plants and the more effective the organic system will be. Remember, proper nutrition is the corner stone of the organic system and compost is it's main building block. Roses love compost so feed it a cup of good rich compost once per month.

"Provide for your roses a constant supply of good rich compost"

Organic Fertilizers

Chemical fertilizers lack bacteria and enzymes essential for soil life and for nutritional exchanges necessary between plant and soil. Compost is the finest organic fertilizer you can use on your roses. However, you will need to add a good organic fertilizer, apprx 1 cup per month per plant. The Organic fertilizer should be a 5-5-5 with Nitrogen no higher then 7. There are many good organic fertilizers on the market today. Read the ingredients. Avoid urea base fertilizers. Find two or more different organic fertilizers and switch between them. Or you can make your own fertilizer.

Making your Own Organic Rose Fertilizer

1 lb New Jersey Greensand
2 lbs Rock Dust
2 lbs Alfalfa meal
2 lbs Nature Meal(from Nitron)
2 lbs Fish meal
2 lbs Seaweed powder like Maxicrop, etc.
2 lbs Earthworm Castings
1/2 lb Epson Salt
1 lb Chelated Iron
1 lb Bone Meal

Mix together, use at 1 cup per plant per month. Water well. See also organic fertilizer chapter for another version of rose fertilizer.

Minerals

Minerals in a fine form such as rock dust, granite dust, decomposed gravel, greensand, soft rock phosphate. Minerals from the ocean such as kelp meal, fish meal, seaweed powder, crushed oyster shells. Minerals from the animal kingdom such as bone meal, feather meal, minerals from animal manure.

NATURAL PEST CONTROL

Using Rock Dust

Rock dust will help to increase the energy level of the soil and in turn will quickly raise the roses and the soils energy levels. This is primarily due to its high calcium levels as well as high iron, and its large selection of trace minerals, which are made immediately available to the soil and plants. To avoid the dust, rock dust can be made into a milk like liquid and sprayed on the leaves. Use only 1 tablespoonful of rock dust and 1 tablespoon of Diatomaceous Earth(garden grade only) per gallon of distilled or solarized water(Stir in a clockwise direction for 1 min then quickly stirring in the opposite direction for an additional min.). Allow to settle for 5 mins. Add 10 drops Superseaweed(TM) or any concentrated liquid seaweed or natural fish emulsion. Strain into sprayer. Spray once per month, or daily, as long as pest infestation occurs.

Step 4

Increasing the Bacteria

Bacteria are natures cooks. They take raw materials(minerals) and eat them, converting the minerals into compounds which are easily assimilated by the plants. The bacteria eat first! We are in essence feeding the soil! Try to use a garden filter to filter out the chlorine in the city water. Chlorine kills bacteria.

Bacterial sources come from any of the following

Animal Manures provide many different types of bacteria needed by the soil. Choose from different types of animal manures. Horse, cattle, sheep, llama, rabbit, are some to name a few. Do not use your dog or cat manure. Do not use any carnivorous animals as their manure could contain harmful parasites. Take a drive out into the country and discover for yourself what resources are available for you in that area. Chicken farms,(egg farms), rabbit farms, dairies, horse farms, etc. all make good sources of these materials. Talk with them and see what they want for it. Sometimes they'll give it to you free! Ask them if they spray any chemicals on the manure. Avoid using any manure that has been recently sprayed. Allow at least 6 months before using. Composting will remove most chemicals.

Other sources of essential bacteria: blood meal, liquid seaweed, fish emulsion, (Nitron A35, Agri-Gro,Shure Crop, Superseaweed, Willard Water, to name a few commercially produced), milk, molasses, aged tree bark, Alfalfa meal, herbs. Learn The Bio-Dynamic system.

Page 78

This will help you to increase your basic understanding of how Mother Nature has put everything together and how it works and best of all, how to utilize it to your benefit.

For bacterial sprays
use any of the following to increase bacterial activity.

Nitron A35:

A Bacteriological activator. Provides enzymes. Helps bring the soil back to life. Use at 1 cup per plant per month. Ok to use along with organic fertilizer but will need to use 1/4 cup fertilizer less, as the Nitron will make more of the fertilizer available. Available from Nitron Industries, 1-800-835-0123.

AgriGrow

Provides many natural bacteria needed by the soil. See foliar spray chapter for more info.

Using a Liquid Seaweed:

Seaweed is full of trace mineral and bacteria. For more information see chapter on making your own superseaweed.

Superseaweed: Provides trace minerals, bacteria. A natural Biodynamic spray. Use 10 drops per gallon. For best results add 10 drops per gallon of filtered water one day before use and allow to sit. Superseaweed can be added to Nitron for even better results results. For more information see chapter on Mamking your own Superseaweed. Available at IG of A 1-800-354-9296.

NATURAL PEST CONTROL

Compost Tea: One of the best bacterial sprays is made from compost tea. Place 1 cup of compost into a panty hose and tie into a ball. Suspend in a glass gallon container filled with filtered water. Allow to sit in sun for 24 hours. Spray in evening. A good program is to add 1 cup Nitron, 10 drops Superseaweed into gallon of compost tea. Allow to sit in sun for 24 hours, then spray on roses.

Milk: An excellent bacterial spray, milk is also a good source of calcium. Milk provides a natural bacteria that prevents and fights various diseases. Milk will kill off fungal diseases while allowing beneficial bacteria to grow. Fill a gallon sprayer to within 3 inches of the top with filtered water or solarized water. Solarized water can be made by filling a gallon glass container. (A good container is the water companies 5 gallon bottle.).

Please note: There are many different products on the markets which can be used in place of the above. We however can only recommend those products which we have used and tested in our organic testing gardens. If you have a product(s) that you think we should know about, please see appendix.

Allow to sit in sun. Shake regularly to remove any chlorine, etc. One or two days will be long enough. Spray on leaves. You can add a liquid seaweed concentrate such as Superseaweed(TM) or MaxiCrop(TM) to increase effectiveness.

Fish Emulsion: Fish emulsion makes a great bacterial spray. Try deodorized fish emulsion for less smell. Avoid fish emulsion which has urea in it.

ALfalfa Tea: The biodynamic farmers knew the benefits of spraying with Alfalfa tea in early spring. Alfalfa is very high in nitrogen(10%), very high in many trace minerals, high in iron and especially high in natural bacteria. To make Alfalfa tea add 5 cups Alfalfa meal into a 5 gallon container of filtered water. Allow to sit for one day. Then add 1/2 cup fish emulsion, 1 cup Nitron A35, 20 drops Superseaweed and 1 cup rock dust. Stir well and allow to sit in sun for another day. Should start to smell just right. Add 1 cup of this liquid to 1/2 gallon of water. Can be poured around base of roses or sprayed on leaves. If spraying, filter before pouring into sprayer.

Step 5

Mulch well

Mulching is very important. Use acid mulch if soil is too alkaline, use alkaline mulch if soil is too acid. Mulching helps retain moisture, keeps nutrition levels high and reduces stress. Proper mulching increases bio-diversity. The more bio-diversity, the greater benefits between living organisms. This reduces any infestation of any one species(like snail control).

Step 6

Disease and Pest Controls

Once you have begun to provide for a proper nutritional environment, you must provide protection until balance has been regained. You can protect your roses from thrips, aphids and other pest attacks thru several different methods:

Planting garlic at base of rose will deter pest attacks. Society garlic is very effective, also. You can also make a garlic oil spray. Purchase garlic oil from your local grocery store. If they don't have garlic oil buy garlic butter. Add 1 tablespoon to a quart water sprayer. Making sure to strain before pouring into sprayer to prevent clogging.

Another method you can use to control pests on your roses is to bury tobacco around the base of your roses. For best results take 1 cup dried tobacco(use only organically grown tobacco such as available from Santa Fe Natural Tobacco Company in Santa FE New Mexico), 1/2 cup garlic powder and mix into this 1 cup compost. Bury at base of rose by turning over into the soil. A better method and one which I prefer, is to get a small clay drain pipe(opened at both ends, about 3 inches wide, 12 inches long(smaller for roses, larger for trees)) and bury that at the base of the rose bush or plant, level with the ground. Into this place the mixture TGC+(TM)(equal amounts of organic tobacco, garlic powder, compost and the + stands for trace minerals such as rock dust).

Use 1/2 lb per unit. On top of this place either a pretty rock or a nice clay pot to cover it, or simply mulch over, remembering where it is. This mixture should be replaced once per year or as often as needed.

You can also use pebbles as a mulch to cover it, just remember where it is. For this rose vent to be effective, there should be a drip head (2 gallons per hour)to allow water to pass thru it. The TGC+ will be absorbed by the rose, Ficus etc, and it will kill the thrips, aphids,

NATURAL PEST CONTROL

etc., as they attack the rose leaves(anything that eats these leaves will die!). Tobacco and garlic are both absorbed into the plant, anything that attacks the plant will also get this mixture. The tobacco will kill any insect that attacks it.

The Tobacco is very volatile and will biodegrades within 24 hrs if in liquid form;will last longer if it is in a dust or in its natural leaf state. Do not use tobacco on fruit trees. The garlic will change the taste of the plant confusing the bug.

Tabasco Soap

Another spray formula to use to control thrips, aphids and most pests: add 1 tablespoon tabasco sauce, 1 tablespoon natural soap such as Dr. Bronners Peppermint Soap to a quart water spray bottle. This can be sprayed onto the leaves and around base of plants. It will repel and kill thrips, aphids, etc. You can alternatively add 1 tablespoon garden grade DE. When using DE, you must be careful to avoid breathing or eye contact, due to its abrasiveness. Garden grade DE is perfectly safe(pool grade DE is dangerous) to use but it is a dust and should be handled with care.

Wear a face mask when using and wash with water if it gets in your eye. Don't rub your eyes, remember DE is like thousands of tiny razor blades. Water will wash it away.

DE

Diatomaceous Earth is a natural dust used to control crawling insects, beetles, ants, aphids, spiders, snails, etc. DE can be used in many ways. Can be added to water, strained and then sprayed. Use 10 tablespoons per gallon filtered water. DE can be made into a paste and then painted around the base of the roses to prevent caterpillars, snails, etc., from climbing up to eat the flowers. For more information on DE see chapter on making your own pest control.

Controlling Ants on Roses

The other day I had a caller on my radio show. She had heard my previous remarks to a caller concerning ants and aphids and their relationship. This caller was certain that only some aphids are herded by ants and that most fly from place to place totally indifferent to what the ants have to say about it. It is just the opposite, that most aphids are under the control of their ant masters.

I also have found it to be true that by controlling the ants, you obtain a greater control of the aphids which are attacking your roses, etc. There is a definite rela-

tionship between various insects, in particular aphids, which are found on plants, and ants. The ants control most if not all insect activity on plants which they have 'adopted'.

Plants under the care of the ants are protected by them and are also used as a source of food either for them directly thru the sap or pollen of the plant or indirectly thru the use of aphids and other insects which attack the plant and which in turn are 'milked' by the ants (for their nectar).

Changing the behavior of the ants is a very important factor in controlling many pests in the home and garden as well as controlling the ants themselves. This is developing a line of communication between the ants and yourself. Ants are uniquely positioned in the insect kingdom.

They are intelligent enough to remember. Ant memory works in a very direct way. They are 'programmed by nature to behave in a certain way. Their actions are controlled by certain factors in their environment. Control these factors and you control the ants. Quiet Control is what you seek. It is not neccessary to kill them. For more information on ants see chapter "Dances with Ants".

Getting A Hold

Raise the energy level of the soil; in turn you will have higher energy levels of the rose, vegetable or plant. The higher the energy level of the soil, the healthier the rose, the less stressed and the less pests your roses will have attacking them. Ants respond to inbalance and stress.

Give to Caesar his Due

Feeding the ants will reduce their activity in other parts of their kingdom(our yard and house). Ants like most creatures except man, follow the path of least resistance. Ants are good at this. If an endless food source has been found, they will use it to the benefit of their colony.

Ants will not have to go looking for food as long as they get what they need. The basic idea is to provide for the ants a feeding center, see Ant Cafes(TM)chapter one. Set up at least one Ant Cafe near by. By providing them a food source, we can begin to retrain them to come for food here and to stop looking for food in the kitchen, or on the roses, etc..

This is a simple system that can become your most effective control method against the ants.

NATURAL PEST CONTROL

Using Natural Barriers to keep Ants off

Tangelfoot works well for this. You can add cayenne pepper to increase its effectiveness. Other barriers to keep ants and aphids off are Tabasco sauce and a natural soap sprayed on base and on leaves. Use 1 tablespoon of ea in 1 quart water. A good soap to use is Dr. Bronners Peppermint soap which is available at most health food stores. Any natural soap which has a strong fragrance will work, Peppermint is very effective for this purpose. Natural fungicides used to control Downey Mildew, Rust, BlackSpot, Powdery Mildew and other exotic diseases:

By now you should get the idea that the basis for regaining the health of the roses and other plants is to regain the health of the soil. The key word here is "Exotic". This is very important to understand. "Exotic" diseases can be directly correlated to the lack of "Exotic" trace minerals. Usually it is an absence of, rather then the presence of these exotic minerals that triggers the effect (which is the disease). This is also true for common diseases.

To regain control of any such disease, you must regain the balance of the eco-system that the plant is growing in. Check soil PH levels, Check for excessive salt levels such as boron, chlorine, etc. Check watering habits and equipment. Use a garden filter to filter out the chlorine. Check your composting/fertilizer habits, Switch over slowly to organics. See more info on Dances with Ants chapter.

Natural Methods of Fungus Control

Milk, Garlic(use 1/2 tablespoon crushed garlic per gallon, strain before using), wettable sulfur(follow instructions on label), Baking soda(use 1 teaspoon per gallon water).See Disease control chart for more uses.

A GOOD BIODYNAMIC FORMULA

Here's a very old Biodynamic formula which will help you to control many diseases on your roses, plants etc. Obtain horse, cow, sheep, llama, or rabbit manure. Make sure its not more then 6 months old. You will need only about 1-5 lbs.The amount depends on the number of roses, etc. you want treated. One cup of this manure mixture will make 3 gallons of a liquid spray. One gallon of this spray will take care of 10 full grown roses, etc. Add 1 cup rock dust per every lb of manure you get. Add 1 cup natural clay per every lb of manure. Add 1 cup powdered seaweed per every lb manure used. Place this mixture into a large clay pot and mix well. The clay pot should have a cover for it. You can either bury in ground(in a shady area) or place in your basement if you have one. Allow to sit for 1 month. Take 1 cup of this mixture and place into a panty hose. Place into a 5 gallon glass container such as the one mentioned above.

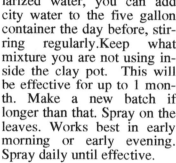

You can buy these from your water company. Allow to sit in the sun for at least 1 hour, three hours is best. Remember to use only filtered or solar water.

If you are going to use solarized water, you can add city water to the five gallon container the day before, stirring regularly.Keep what mixture you are not using inside the clay pot. This will be effective for up to 1 month. Make a new batch if longer than that. Spray on the leaves. Works best in early morning or early evening. Spray daily until effective.

Using Molasses

You can use molasses to control many diseases on your roses, etc. Here's a good formula to use. To one gallon filtered or solarized water add 1 cup unsulphured molasses. To this add a dash of soap such as Dr. Bronners Peppermint soap(1 tablespoon will do as it is a concentrate), any natural soap will do, experiment! Spray on leaves. This formula only works on roses which are already being fed properly and have a good alive compost/soil to rely upon. The molasses provide a special bacteria as well as sugars in a form available to the plants.

Bordeaux Mixture

Bordeaux Mixture works well against most mildewsand fungus such as Powdery Mildew and Rust. You can make your own Bordeaux mixture by add three ounces of copper sulfate(bluestone) to 3 gallons filtered water and dissolve well. Add 5 ounces of

NATURAL PEST CONTROL

hydrated lime and again mix well. Ready to use at 1 part mixture to 1 part additional water. Can be used at full strength for bad infestations.

FINE HORTICULTURAL OIL

Using Sun Sprays Ultra Fine Horticultural oil will smother the rust spores and it will reduce them during the growing season. Fine Horticultural Oil can be used year round without burning the plants. If used during the early spring , it will help to control and reduce many fungi. Available thru Gardens Alive!

BIOLOGICALS

Biologicals are used to control many varieties of rose attacking caterpillars: BT, MVP, Neem(available from Gardens Alive, etc.). Use Tangelfoot(available from most garden centers) and place a line around the base of the plants. It will keep caterpillars(and snails) from climbing up. DE made into a paste and painted around base of rose will work also.

SOAP

Using Soap will control spider mites,aphids, ants, thrips and many other rose pests. Use 5 tablespoons of Dr. Bronners Peppermint soap, or any other natural soap, per gallon filtered water. Safer Insecticidal soap is also excellent to use.

Proper pruning is very important and can only be learned from experience.

NATURAL PEST CONTROL

Chapter 13
Organic Tree Care

About Trees

Trees are very important to this planets ecosystem. They are essential to the earths recycling system. They provide the very air we need to live. They control the water cycles. They provide homes to many different species. They have been called the earths shock absorber. They assimilate carbon into oxygen, they clean up the air(improve air quality) and bring water and minerals up to the surface. Reduce water run off. Urban forestry helps clean up cities and raise the quality of life, promoting interaction between urban dwellers and their environment. Trees also reduce heat and glare.

Why Feed them Chemicals ?

The answer is simple, trees cannot use chemicals in their system! Trees must be fed naturally and in accord with how the natural system works. Remember trees have been around for a long time.

Proper Tree Care

How to feed them then? Chemical fertilizers damage the soils ecosystem, cause stress and lack the essential bacteria and minerals. Trees require certain conditions for proper growth and good health. Proper site selection is very important.It is important to understand what these conditions are. Proper tree selection for the environment it will be in is also important.

Before planting amend the soil with good rich compost and lots of humus. Provide regular waterings. Provide regular composting.

Mulch as needed

Most trees will mulch themselves so don't rake away their leaf droppings. Do not use any chemical fertilizer or pesticides. Do not top trees! Go Native when ever possible. Hire only skilled workers. Remember, it cost less to keep a healthy tree then to replace it. Save a tree whenever possible. Don't cut it down, plan around and include the tree in your planning!

Trees need Water

Make sure that you provide adequate watering on a regular basis. A drip system or soaker will help. Filter the water. Use tree vents for deep watering if possible. Use a garden filter for this. The proper selection of trees is important for this purpose.

Trees Love Compost

Compost is one of the few things you can feed trees with. Local compost rich with trace minerals, composted animal manure and lots of humus are best. Feed yearly if possible.

Never feed trees High Nitrogen

High nitrogen causes stress which is the main cause for pest or disease attack. Trees get their nitrogen from the air, from the compost and from the natural bacteria found in the soil.

Tree Vents

A tree vent is a clay pipe that is placed under a tree. Four tree vents per tree equally spaced about 2-4 feet away from the truck (depending on size or age of tree). I use clay drain pipes that are 3 inches wide and 12 inches long, long enough for our purpose. For a very old tree try 24 inches long (Two 12 inch tree vents stacked on top of each other). Run a drip line to water the vents. A 2 gallon per hour head in each vent will do. Filter water. Water at least once per month or as needed.

Inside the tree vent place compost, rock dust, etc. Also in the tree vents you can place the following mixture if the tree is being attacked by pests or diseases.

The Tree Vent Formula
Makes enough for four tree vents
1 cup carbonated water
1 gallon Nitron
8 oz   Superseaweed
1 gallon Shure Crop
4 lbs Compost
2 lbs organic smoking tobacco
2 lbs Rock dust
1/2 lb Garlic

NORWAY MAPLE

NATURAL PEST CONTROL

Blend together, add 2 lbs into each tree vent. A coffee cans worth is apprx 1 lb. Water well. Add 1 cup Nitron to each tree vent, add 1 cup Shure crop to each tree vent, add 10 drops Superseaweed to each tree vent. Add 1/4 cup carbonated water into each tree vent. Water again. Do not allow water to overflow. Water slowly to fill. Repeat the Nitron, Shure crop, and Superseaweed and carbonated water again in two weeks. Then again in a month. Repeat the tree vent formula in two months time, then apply yearly.

Placing the Tree Vents

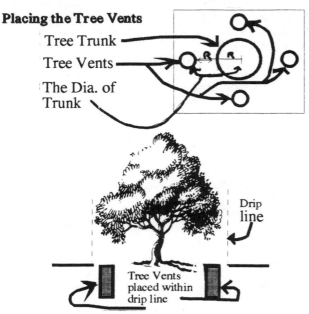

Tree Trunk

Tree Vents

The Dia. of Trunk

Drip line

Tree Vents placed within drip line

Make sure drip system is working. Do not allow to dry. You can use tobacco dust or buy organic tobacco from Santa Fe Natural Tobacco Co in Santa Fe New Mexico. Tree vents can be used for all trees placing only compost as a food source. This provides for deep watering, encourages root development, provides oxygen to the soil. The tree vents are an essential in pest control since the tobacco will kill any pests that are attacking it. The Garlic will prevent bacterial diseases from spreading. The carbonated water provides trees with a source of carbon dioxide.

Natural Spraying

Trees love to be sprayed with many natural products. Seaweed is very good for them as is fish emulsion (without urea). Making a milk out of rock dust is very good for them. This provides them with many natural trace minerals, calcium, iron, magnesium etc. Nitron is good to spray them with. Shur Crop is a good bacterial spray for trees.

Superseaweed is an excellent overall seaweed concentrate to use. Provides trace minerals and bacteria. Use only 10 drops per gallon. Follow instructions on

all labels. There are many natural products on the market these days. Choose carefully what you will use for the trees, ask questions, avoid using chemicals. Follow the Law of the little bit when spraying. If you are un certain about the amount to spray, spray less, more often. When working with trees give them time to respond. Sometimes results won't happen until the folowing season!

Natural Pest Control Methods

Healthy trees will not be affected by insects and diseases as much as sick trees are. That is your first line of defense. However what to do until the tree gets healthier is very important.

Plant pests and diseases play an important role in controlling the tree population. Not all tree pests are harmful to the tree some are beneficial.

Learn to identify the symptoms of plant problems, experience will teach you.

Here is a listing of various possible insects and diseases which attack trees. Natural Controls, Causes and associated Nutritional Deficiencies(in order of cause) are listed at end of each.

Pests

Aphids attack by sucking tree juices, attract many other insects and cause fungus to spread. Control is soap and water. Add 2 tablespoons of any natural soap or insecticidal soap, add 2 tablespoons tabasco sauce per gallon water. Spray tree.

Remember that aphids are attacking the tree because the tree is stressed,is it near a lawn? High nitrogen causes stress. Usually means lack of water, too much

NATURAL PEST CONTROL

water or a lack of proper nutrition. Needs calcium, iron. Provide compost. Other stresses can come from environmental causes.

Leaf feeding Beetles can be controlled by using tree vents with 1 lb TGC+(Tobacco, Garlic, Compost, Rock dust). This will take time to be effective, also depends on how well root system works. You can use beetle traps, nicotine sulfate, sabadilla, ryania, DE , rotenone is effective also. A fine horticultural spray will kill beetle larvae, or you can use 1 Oz Coconut oil mixed with 1 oz soap, pour into 1 gallon water. If you use beneficial nematodes, or spined bean beetle, spray only seaweed as using any of the above will also kill the beneficials.Key in to proper nutrition.Use compost!

For Borers use tree vents to control. Use tobacco pushed into holes made by borers. Trees can be sprayed with DE ,soap and seaweed. Fill tree vents with tobacco and garlic. Water well. Provide trace minerals in a compost form.

For Leaf feeding Caterpillars use BT or MVP add soap and molasses to mixture. Trace minerals.

For Leafminers use BT or MVP, add soap and molasses.Stress due to lack of trace minerals.

For Nematodes use beneficial nematodes or use clandosan.

For Plant Bugs use soaps, DE.

For Scale use soaps. Dont over do it as too much soap will burn. Scales are usually connected to the presence of ants. Use Tangelfoot to keep ants off the tree.

To control Spider Mites use a mixture of soap and DE and Tabasco sauce and a liquid seaweed or Nitron.

To control Tent forming Caterpillars use BT or MVP and soap.

Diseases

To control Cankers promote the natural bacteria in the soil. Spray with compost tea and Nitron or Superseaweed. Use lots of compost rich in rock dust. Use an acid mulch.

A Chlorosis tree needs minerals not just iron. Provide rock dust once or twice per year. Use lots of compost and mulch as needed. Use tree vents to feed compost to trees.

To help fight tree Decay provide tree with rock dust, compost and mulch as needed. Spray with compost tea and Superseaweed or Nitron.

To control Fireblight provide lots of compost and mulch, spray with compost tea and Nitron or Superseaweed. Sulfur will help here but use a small amount.See chapter on making your own foliar spray chapter for how to make compost tea.

Fungal diseases are caused by dead soil. This is taken care of by applying compost and mulch yearly. Spray with compost tea, Nitron or Superseaweed.

To control Gall formers use soap and DE mixed with water. Nutrition is very important.

To Control Viruses provide lots of compost and mulch. Spray with compost tea, add a tsp. of sulphur and a tsp. of garlic. Also a good spraying of Rock dust milk and Superseaweed, Nitron or any type of liquid seaweed.

Environmental Damages

This is damage due to an environmental reason such as weather variations, chemicals in the air, soil, and water. Other damages include earthquakes, flooding, fires, etc. There is not much that you can do about these except keep the tree as healthy as possible.

Herbicide injuries. If herbicides are used flush with lots of water. Provide compost and mulch. Spray a seaweed like Superseaweed or use Nitron.

For Pesticide injuries flush area with water. Provide compost and mulch. Spray with a liquid seaweed mixture.

For Leaf scorch make sure tree is getting plenty of water. Always filter water. Use a drip system to water tree vents.

To fight Drought injuries use compost and mulch. Use a drip system to provide water. Use tree vents to allow for deep watering and feeding.

For Salt toxicity City water contain chlorine, a salt. Use filtered water only on trees. Apply compost and mulch. Avoid fertilizers as they are all salt based.

Winter injuries Most trees will recover from winter injuries if strong(hint). Feed well with compost twice per year if possible, spray with rock dust milk or seaweed.

NATURAL PEST CONTROL

NATURAL PEST CONTROL

Some Rules for Spraying

A 30 gallon sprayer is best to spray trees. Spray trees either early in the morning or late in the afternoon. Avoid spraying if temperatures are above 90 degrees F. Never spray one particular element, always spray a natural blend of sources.Always deep water trees before spraying.

For more information see making your own foliar spray chapter.

Sugar and Trees

Using molasses as a spray, you are providing trees with a source of energy. Using compost rich in phosphate allows for greater sugar production in trees. Raising the sugar content of trees allows for greater mineral absorbtion. Use 1 quart per 30 gallon.Add 1 quart of Milk to increase bacterial activity. Calcium is an important source of energy for trees.

H2O2(Hydrogen Peroxide) can be used for many different fungal diseases in trees. Use 1 quart per 30 gallon.

Using **Vinegar** helps to reduce PH levels. Use 1 quart per 30 gallon.

Carbonated water increases carbon dioxide to trees. Use 1 gallon per 30 gallon.

DE(Garden Grade Diatomaceous Earth) Use 1/2 lb per 30 gallon sprayer. Add 1 lb to bucket, slowly add water to fill. Dissolve DE. Allow to sit then strain into a 30 gallon sprayer.See Using DE in pest control chapter.

Rock Dust. When using rock dust use 1/2 lb per 30 gallon sprayer. Make the same way you made the DE. Add 1/2 lb to bucket. Slowly fill with water and stir to dissolve. Allow to sit a few minutes and strain into 30 gallon sprayer.

RULES FOR NATURAL TREE CARE

.Never feed trees a chemical fertilizer.

Trees require regular watering. Use a drip system whenever possible. Avoid water trunk of tree. Trees love slow deep watering.

Always use filtered or solarized or well water. Never use straight city water as it may contain chlorine or some other chemical that can poison the tree. See garden Filter Page.

Allow time for trees to Heal. The bigger the tree, the longer it will take to heal.

Raise the energy level of the tree thru the addition of compost rich in trace minerals and bacteria. Trees love rock Dust!

Know the law of cause and effect. Control the cause and you control the effects. Trees with high energy levels are not attacked by pests.

Don't cut down trees.

Learn your trees. Plant only trees that will grow in your area. Give your favorite tree a name.

Protect trees from dogs. Dog urine is very acidic and will injure the tree's root system.

Cultivate the soil regularly. The soil will get packed so tha water and air may not reach the roots.

.Mulch. Provide plenty of mulch.

Protect tree's from people!

Do not poison the soil. Avoid spilling gasoline, paint, salt etc...

Remove any tight wire supports as this will strangle the tree as it grows.

Plant ground covers. They make a great living mulch which conserves water

BOUNDARY ELM, 1847.

NATURAL TREE CARE

Rules of Natural Tree Care

Rule #1   Never feed trees a chemical fertilizer.

Rule #2   Trees require regular watering. Use a drip system whenever possible
Trees need slow deep watering.

Rule #3   Always use filtered or solarized or well water.Never use straight city water as it may contain chlorine or ? which maybe harmful to the tree.

Rule #4   Allow time for trees to heal. The bigger the tree the longer it will take to heal.

Rule #5   Raise the energy level of the tree thru the addition of compost and trace minerals in the form of rock dust. For best results add the rock dust to the compost.

Rule #6   Know the law of cause and effect. Control the cause and you control the effects. Trees with high energy levels are not attacked by pests. If an attack occurs they are repelled better by healthier trees then weak trees.

Rule #7   Don't cut down a tree unless you have to. Buy a live Christmas tree and plant it after Christmas or donate it to a local club or school.

Rule #8   Learn your trees. Plant only trees that will grow in your area.

Rule #9   Protect trees from dogs. Dog urine is very acidic and will injure the tree's root system.

Rule #10   Cultivate the soil regularly. The soil will get packed so that water and air may not reach the roots.

Rule #11 Mulch. Mulch protects the compost, reduces weeds, retains moisture and prevents the soil from hardening.

Rule #12   Protect tree from people. Do not allow people to chain bikes, to cut into the bark or lean heavy objects against the tree. Trees do not like being painted either!

Rule #13 Do not poison the soil. Avoid spilling gasoline, paint,salt, oil, detergents at base of tree.If this happens water well but first try to remove as much of the affected soil as possible.

Rule #14 Remove any tight wire supports as this will strangle the tree as it grows.
Rule #15 Plant ground covers. They make a great living mulch and add beauty also.

Rule #16 Do not raise the soil level around the base of the tree.
Rule #17 Use tree guards or fences when ever possible.

NATURAL PEST CONTROL

Chapter # 14
Making your own Organic Foliar Sprays

The word Foliar means leaves. All plants have leaves of one form or another and when you feed them through these leaves you are foliar spraying them. The idea is to provide plants with important trace minerals, bacteria, enzymes, and elements necessary for a healthy plant to grow.

This Chapter will strictly discuss the application of nutritional foliar sprayings and not the use of foliar sprayings concerning pest controls. That subject is covered in the chapter on Pest controls.

About Seaweed (Kelp)

One of the best foliar sprays comes from Seaweed or Kelp. Seaweed has been used for centuries as a fertilizer to grow mankind's food. The ancients all over the world knew the value of the ocean and her importance to their survival. Seaweed contains many trace minerals, enzymes, and bacteria needed for healthy soil and healthy plants.

Rock Dust Milk

Rock dust can be made into a liquid for spraying, see chapter on rock dust. This is an excellent way to supply trace minerals especially calcium and iron to soil and plants.

To better understand the many ways to use this valuable resource, we must look to our past.

A TRIP INTO THE PAST

ISLAND LIVING

Anyone who has ever lived on an island quickly learns the value of the ocean. Being surrounded by water, island living relies on the resources of the ocean. Mankind quickly developed special relationships between themselves and the watery world. Food of course was found in this world. Also found here were the ingredients to be help man grow food in the barren island soil. Seaweed was dried and used as a fertilizer, along with anything and everything that came from the ocean.

American Indians

The American Indian learned the secrets of the world and passed this knowledge to their children, and children's children. This knowledge was in part the power that came from the ocean which could be harnessed into making plants grow better and produce more. Their method of burying fish for the growing of their corn is well known. This knowledge is still used by many people.

Peru

The barren land would not give to them unless they gave to her. Thus the relationship between the peoples of Peru and the oceans would become a very important one. Peru has a very ancient knowledge 'pool' from which to choose. It is from this 'pool' given to them from ancient times, that they learned how to use the soil. By giving it the richness of the ocean the barren land would give untold food back in return.

Central America

The Mayan, and their precursor's the Olmec, flourished for a millennium along Mexico's Gulf coast. The levels to which they brought agriculture and the special relationships they developed with the ocean and the land bear a closer look. To greater understand the reasons why this culture flourished completely in-

NATURAL PEST CONTROL

tertwined with the ocean and the earth around them, we must look at their relations with those around them.

Even though they were not travelers over water, they traded with those that did and thus were able to extend their ability to control the Earth's production of valuable food. Their land was well suited for food production and they quickly developed special food producing areas which were used completely for this purpose. They developed special "Gardeners" who did nothing but work on these fields. They knew the importance of the ocean's nutritional values.

They dried their own seaweed, as well as various ocean creatures such as fish, squid, etc., which they often traded with the fishermen of nearby Honduras, Nicaragua, El Salvador, and even little Belize. It was during the Classic Mayan cultural period (A.D. 300 to 900) that the Mayans developed the art of agriculture to a high degree.

Europe

Europeans were one of the earliest cultures to formally recognize the importance of the ocean to their agricultural practices. This was due largely to the masses of people that needed to feed themselves and to their ability to learn from others.

They used the oceans largesse to increase their production along with using animal manure (of which they had a lot), which formed the basis for their agricultural production up until the chemical revolution in the early 1900's.

These folks also learned fast from other cultures, as they were able seaman and traveled around the world, and would see many wonderful things and this knowledge would return to their own lands, where it quickly would spread like fire upon the land.

China

The mother land has known the secrets of the ocean since time itself began. China is old and so is her knowledge of the ocean. All of the countries of this area knew the ocean- India, Japan; all have learned to rely on the ocean and her bounty.

The ocean is very rich in trace minerals, enzymes, and bacteria as well as small amounts of nitrogen, phosphorus and potassium. This special mixture, when applied to plants/soil, has the ability to release locked up minerals; improve the cation-exchange in the soil, and has chelation abilities.

Seaweed is very high in Potash as well as Magnesium and other more exotic trace minerals such as Boron, Barium, Chromium, Lead, Lithium, Nickel, Rubidium, Silver, Strontium, Tin, Zinc, and even traces of Arsenic, Copper, Cobalt, Molybdenum and Vanadium; all of which are important to proper plant growth as well as proper human growth(over 70 are needed).

All the elements that compose this earth can be found in the ocean (if you look hard enough). Seaweed is important for this reason as a trace mineral source in making compost. It is no wonder that the origin of life was in the ocean.

Here is a partcial listing of various seaweed products on the market to date. I only recommend those products which I have used and tested. If you wish your product to be listed please see appendix "IGA membership info".

AGRI-GRO

Invented by Dr. Joe C.Spruill. Phd. Biochemistry. A biological complex derived from natural compounds, processed through extraction completed by fermentation. A plant and bacterial stimulant, contains stabilizing nitrogen-fixing bacteria, trace minerals, and humic acid.

Designed to improve chemical and fertilizer effi-

NATURAL PEST CONTROL

ciency, improves natural sugar levels of plants, improves tilth and water capacity, designed to reduce cost of production, reduce insect and disease problems and reduce soil compaction and also to reduce nematodes and salt buildup in soil.

AGRIKELP

A good liquid seaweed . Comes from Australia. Very pure and free from heavy metals. Available from : The Continental Shelf

FOLIAGRO

A powder extract from ocean micro-algae (Ascophylum Nodosum). Easy to use. Excellent for making your own Superseaweed!! Available from: Gardener's Supply

Kelp Extract(Algrow Brand)

This is a special kelp extract designed for foliar sprayings, is made from cold processed Norwegian Ascophythllum Seaweed. Cold processing preserves higher levels of minerals and growth hormones than is found in other kelp extracts. This kelp extract is one of the most potent extracts available on the market today! A definite on your making your own Superseaweed list!! Available from: Peaceful Valley.

Liquid Seaweed Concentrate

An Ascophylum Nosodum Seaweed Extract, rich in trace minerals and growth hormones. Available from : North Country Organics

Maxicrop

Maxicrop is an excellent kelp extract containing over 70 trace minerals, growth hormones, cytokinins, auxins, vitamins, and enzymes. World famous as a foliar feeder and plant stimulant. Contains 1% N, 0%P, and

3%K plus elements from the ocean. Comes as a water soluble powder which mixes easily with water. Use 1/2 teaspoon per gallon or also available as a Liquid concentrate. A must in making your own superseaweed! Available from: Peaceful Valley and from Mellinger's

Nitro/Max

Provides more efficient use of residual and applied nitrogen. A Non-toxic, non-petroleum based liquid concentrate designed to increase the natural nitrogen cycle. Available at J&J Agri-Products. See directory.

Norwegian Kelp Meal (Algit brand)

One of the best sources of trace minerals. Contains over 70 trace minerals, plus plant growth regulator's, vitamins, hormones and enzymes!! Harvested off the coast of Norway, the kelp meal is washed, making it the lowest in sodium of any kelp meal. Can be made into a tea and strained and sprayed on leaves of plants or best used as a soil additive. Right up there on the top ten list for making your own Superseaweed! Available from: Peaceful Valley

Sea Crop/Sea Mix

One of the best liquid seaweed's around. A must for making your own Superseaweed!(Seamix is fish emulsion and liquid kelp, highly recommended!!) Available from: Nitron

Shur Crop

Uses Icelandic Kelp Meal as a base. The nutrients are extracted by using a slow fermentation process at a low temperature. This a very biologically active

NATURAL PEST CONTROL

"kelp beer" and must be kept cool. Ingredients:Icelandic Kelp meal, Leached shale, Humates, Molasses, Coconut Oil, Pine Extracts, Onion and Garlic, Diastatic Malt, Yeasts and lots of other goodies!! A great foliar spray for plants. Promotes quick recovery. Works great with Superseaweed!! A must to make your own Superseaweed!

Available from: Peaceful Valley; Manufactured by Hi-Bar Ltd, P.O.Box Kusel Road, Oroville, Ca., 95965. call 916-534-7603

SM 3

A Liquid concentrate of seaweed. From England. Great as a soil activator, as a foliar spray, or even as a seed inoculant! Can be used for making your own Superseaweed. Available from: Mellinger's

Soluble Sea Extract

A liquid seaweed concentrate without the water. The dry extract mixes easily with water. Available from: North Country Organics

Other products from the ocean

Dehydrated Fish Powder: Mermaid brand is a cold processed, water soluble fish powder which makes into an excellent foliar feeder. 10-1-1 analysis. Available from: Peaceful Valley.

Fish Emulsion

My favorite is Atlas Brand, available at most nurseries. Homegrown Brand 5-2-2. A great fish emulsion containing both macro elements and trace minerals. This brand is available from: Peaceful Valley, Alaska Fish Emulsion 5-1-1.

Alaska Brand More Bloom

0-10-10. Alaska Bloom is a light fish base concentration high in phosphorus and potassium that stimulates budding and blooming, to produce more fruits, vegetables, and bigger flowers. Both are available from: Mellinger's

Fish Meal

Homegrown Brand is a good source of all major trace minerals and major elements. Available from: Peaceful Valley.

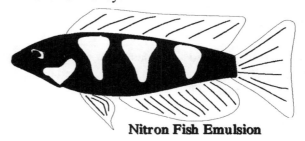

Nitron Fish Emulsion

A high Quality Non-Urea bases Fermented Fish emulsion. 5-0-0 A great Foliar feeder. Available from Nitron. The best on the market! Very little ordor too!

Odorless Liquid Fish(4-8-4)

A refined fish emulsion that has no odor Available from:North Country Organics.

SeaLife Meal:
A great kelp meal based product. Available from Nitron.

Superseaweed

Developed by The Invisible Gardener. The ocean has always been a source of nourishment for the plant and animal kingdoms. The Oceans provide us with all the trace minerals, enzymes and bacteria necessary for healthy human growth. Plants need to have the same nutrients we do. Ingredients: SUPERSEAWEED is a blend of over five liquid seaweed's from around the world, selected especially for their purity (no heavy metals, toxin's, etc.). NOTHING TO POLLUTE OURSELVES OR THE ENVIRONMENT EXISTS IN THIS PRODUCT. SUPERSEAWEED is a wetting agent. It makes water "wetter",

NATURAL PEST CONTROL

allowing plants to absorb more nutrients.

Because your plants will be using their available fertilizer to capacity, you will be reducing wasted, or unabsorbed fertilizer. Therefore, only 1/2 to 1/3 of your present dose of fertilizer will be necessary. In itself, SUPERSEAWEED IS NOT A FERTILIZER. SUPERSEAWEED is a special "biological activator"... by making nutrients more available to your plants, they'll become stronger and healthier, reducing plant stress, pest problems and fertilizer costs. Plants last longer and look better.

SUPERSEAWEED compliments both chemical and organic fertilizers. I, of course recommend using organic fertilizers whenever possible. Use SUPER-SEAWEED everywhere; flower and vegetable gardens, houseplants, fruit trees, hydroponic systems; everywhere!

INSTRUCTIONS FOR USE

Do not shake! Turn bottle upside down (it comes with its own dropper). Squeeze one drop at a time. Use 5 drops per/gal. for all plants. Use once per week. SEEDS: 1 drop per/gal; soak over-night to promote growth. COMPOST STARTER: 10 drops per/gal. per cubic yard; to help make rich humus. GREENHOUSE: 5 drops/gal.twice per week; use with inline feeder to spray leaves.

VEGETABLES

10 drops/gal., once per week; use only 1/2 usual fertilizer. LAWNS: 10 drops/gal. Excellent for orchids, roses! (5 drops/gal.) **Available from:The Invisible Gardener. See address in Resources Directory.**

Making Your Own ' Superseaweed(tm)'

Here is a formula that will help you to make your own "Superseaweed". This is not the same SS that I make and sell but it will give you great results:

4 part Nitron
4 part Nitro/Max
4 part AGRI-GRO
2 part Maxicrop (made from powder)
2 part Shure Crop

2 part Sea Crop
1 part Algrow brand Kelp
1 part Norwegian Kelp Meal (Algit brand) (made from powder)
1 part fish emulsion (optional)
1/2 part hydrogen peroxide
1/6 part wetting agent(Shaklee's Basic H or Dr Bronner's)

An example formula
1 quart Nitron Formula
1 quart Nitr/Max
1 quart Maxicrop(made from powder)
1 quart Shure Crop
1 quart Sea Crop
1 quart Algrow
1/2 quart Norwegian Kelp liquid concentrate.
1/2 quart fish emulsion
10 oz Hydrogen Peroxide
1 oz Shaklee's Basic H (wetting agent)

Makes 7 quarts total of your very own "Superseaweed"!! Add 20 drops per gallon for foliar sprayings, as well as waterings of houseplants, etc. Use once per week.

For sick plants, trees, use 1 cup per gallon, spray once per week, until plant recovers, then once per month. For lawns use 5 cups per gallon for sprayings. The idea in making your own Superseaweed is that in blending the different liquid seaweed's, you would end up with greater amounts of different trace minerals, bacteria, etc. So when following the above mixtures, please feel free to try the various different types of seaweed's on the market and see which works best for you and your particular situation. Some seaweed's are better for fruit trees then others, as some liquid seaweed's are better at warding off

NATURAL PEST CONTROL

Making your own trace mineral sprays

10 oz Epson salts: Magnesium sulfate
10 oz Manganese 35% Nutra-spray
10 Oz Sodium Molybate contains 40% Mn.
10 Oz Granular Zinc Sulfate 36%zinc, 18% Sulfur.
10 Oz Boric acid: Boron
1 lb Iron: Hampene (contains 13% Iron)
1 lb Greensand
1 lb rock dust
1 lb Bone meal optional
1 lb Bay Crab Meal
1 lb powdered seaweed

Mix together, use at rate of 1 cup per 1 gallon water. A slurry can be made or you can, place a cup of the mixture into panty hose and allow to soak in 1 gallon water for 1 hour, squeeze into water then spray onto leaves in early AM if possible, or early afternoon. Apply at the rate of once per month as a foliar feeder. Best applied during winter months(if possible).

OK to use fish emulsion or to add more liquid seaweed (following manufacturers instructions), or to add Nitron or Shur Crop (again following manufacturers instructions), during summer months only, as you don't want to encourage new growth during winter months.

Making Your Own Bacterial Sprays

Nitron

Enzymes is what Nitron is all about. Enzymes is what life is all about, too. Without enzymes there would be no life on this planet. Enzymes are positively charged ions. They perform many tasks.

They are important in the detoxification of the soil; soil that is dead from the chemicals poured on them. They help to soften the soil and allow deeper root systems. Enzymes allow water to penetrate deeper into the soil. Enzymes are important to plants in that they help to release minerals, nutrients that are locked up in the soil. This is a must to include when making your own Superseaweed.

Agri-Gro

A natural Solution containing living organisms, enzymes, azotobacter, bacillus, clostridium, humic acid and trace minerals. Agri-Gro stimulates plant growth, promotes health, maintains the beauty of the plant,loosens the soil, is totally " chemical free " and therefore is safe around humans and pets.

Shure Crop

This product is very rich in bacteria and enzymes.
see spraying chapter

Bloodmeal(optional)

The bacteria in bloodmeal is very important to plant growth. Place 1 cup bloodmeal into a panty hose and place in 5 gallons water. Allow to sit for 1 hour in sun. Spray in evenings.

Compost Tea

This is one of the most important bacterial sprays you can ever learn to make!! Compost is rich in just the right types of bacteria necessary to proper plant growth. Place 1 cup compost into a panty hose and place into 5 gallon water. Place 4 hours in sun. Spray in early am or late afternoon. This is an excellent cure for many bacteriological problems such as fungus,etc. See Compost Tea in Appendix.

NATURAL PEST CONTROL

HUMIC ACID

Comes from Leonardite Ore, millions of years old. Rich in humic acid which helps to break down Organic matter. Available from Nitron.

A special wetting agent:

Here's a little secret.
Use Shaklee's Basic H(concentrated soap) as a wetting agent for your sprays. You only need to add 5 drops per gallon. Works great!!! Can't get Shaklee's? Use Dr. Bronner's Soaps or Amways LOC.

Oxygen

Plants need oxygen too. Try this.....
Go to the drug store and buy the 2% Hydrogen Peroxide. Add 1 oz per gallon to your spraying mixture. Your plants will love it. This oxygen will allow greater absorbtion of nutrients into their system.

Developing a Spraying Program

Here is a spraying schedule which you will learn to adapt to your own use. Follow it and learn from it.

WINTER

Winter months spray trace minerals on bark of trees, onto leaves of plants. Spray once per month. Add very little bacterials to spraying's as they will not be useful in the winter months even in warm climates. Use formula #1 for this. Note : this is for Southern areas with warm Winter. You don't spray during snowing months.

SPRING

Introduce bacteria as early as possible. Apply compost tea in spraying's. Use Nitron or Shure crop or AgriGrow to boost bacteria counts along with the compost tea.

Add 1 lb rock dust/5 lbs Maxi Crop to 5 gallons water, or use Superseaweed or your own Superseaweed. Use gallon of this mixture per 5 gallons of water. This is the spraying mixture you will use. If you can't make your own Superseaweed, you can obtain the real SuperSeaweed from The Invisible Gardener/ARBICO,Eco-Source.

SUMMER

To 5 gallons water add the following:
1/2 cup rock dust mixture (see above)
2 cups seaweed mixture(see above).
compost tea bag
1 cup fish emulsion(optional)

FALL

To 5 gallons water add:
1 cup fish emulsion
1 cup AgriGrow
1 cup Nitron
1 cup Shure Crop
20 drops Superseaweed or
1 cup of your own liquid seaweed mixture
1 cup compost tea
1 lb rock dust

Spraying Tips and Hints

Don't over do it! A little bit goes a long way. Learn what you can mix and cannot mix; start your own mix, no mix list. Remember what works. Keep notes.Tell others about your successes.

Not Listed?
If your product is not listed please see Appendix on Getting Listed.

GARDEN ENGINE.

PRIMORDIAL SOUP
"The Bacteria Eat First"

Organic Gardening depends on having a living soil. The process of assimalation, adsorption and growth in a living system is dependent on bacteria. Bacteria forms the key corner stone for the organic process to happen. I learned early on in my childhood training (my mother and grand mother are both organic gardeners and taught me all about bacteria!),that bacteria is a very useful tool Organic Gardening.: from Compost Tea to producing the various Bio-dynamic sprays.

SuperSeaWeed which is an invention of mine, is a result of the many years of experimentation and assimalation of my mother's and grand mother's knowledge. I like to call **SuperSeaWeed** a *"micro-biological activator"*. In simple terms, it is a special blend of five different types of deep ocean seaweed, chosen for its purity and for its special bacterias. Each seaweed , being from a different part of the world, contains its own special bacteria and its own special minerals. Blending these 5 different liquid seaweeds produces a complete bacterial and mineral "soup". To this mixture I have added rock dust, which increases the seaweeds mineral content. The rock dust is an excellent food source for the bacteria found in the liquid seaweed. I also add activated the wallard water following Bio-dynamic principles. Finally a liquid called Nitron A-35(TM) is added. Nitron A-35(TM) is a special enzyme. More on Nitron A-35(TM) is available in other parts of this book, see index.

When using SuperSeaWeed only 5 drops per gallon is required for regular feeding. However you might want to feed your plants more. I suggest that no more then a dropper full per quart. **SuperSeaWeed** is not a fertilizer and was not meant to take its place. Instead it provides plants a "vitamin pill" effect.

SuperSeaweed also helps to provide needed bacteria and enzymes when used with every watering. Superseaweed allows the soil and plants biological systems to begin to work. Plants and soil require regular composting and applications of mulch to insure a complete and balanced system.

Because of **SuperSeaweeds** biological and mineral contents, it is also an excellent compost activator and can be added to the filtered water sprinkled on the compost, in between layers. Rock dust is also excellent for this purpose. In both cases a small amount is all that is required. Since **Superseaweed** also contains rock dust you are insured against major trace mineral deficiencies as well helping your compost pile to heat up by providing needed bacteria.

Here is a formula you can use **Superseaweed** in(for more information please read chapter "Making Your Own Foliar Sprays":

Household Plant Formula:

When feeding houseplants use only filtered or spring water, add 5 to 10 drops **Superseaweed**, depending on the health of the plant. If the plant is sick and it is the first time you are applying **Superseaweed** then use 1 teaspoon per first gallon with 10 drops per watering for the first months application. Don't forget to add good rich compost to the plants container. Never feed plants a chemical fertilizer! There are many companies which sell many different natural fertilizers. A good example is Nitron Industries, Inc., An IG of A member, Nitron has a great catalog of products. See Nitron in Resources.

To Order **SuperSeaWeed(TM)** see order form. If your company has a natural product that you would like our members to know about you can join IG of A by sending $100 for commericial membership application on your product(s).

NATURAL PEST CONTROL

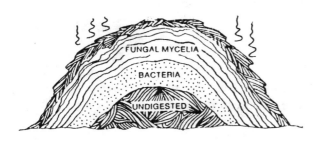

Chapter #15
The Making Of Super Compost

Compost it!

The wheel of life just keeps on turning. From Death comes new Life. The art of making compost is really a science, natures science. As a healing ointment for the soil, Compost is the result of natures dedication to balance. Through watching and learning about the way nature makes compost, we can begin to understand our very own lives and why things effect us the way they do. From Sir Howard to today, the art of composting has taken great steps towards a better understanding of the real process which goes on inside a compost pile.

Purpose of Composting.

There are two major reasons for making compost

To decompose materials down to a more workable and acceptable form

To combine materials in a favorable manner so as to increase nutritional levels as well as increase beneficial bacteria activity, which when added to the soil will increase soil structure.

How Composting works.

Bacterial action in a compost pile is the basis for proper composting. It breaks down the various organic matter, converting it into compost while giving off heat and steam in the process. It also makes available trace minerals to the plants.

Always provide plenty of oxygen (by stirring pile, aerate it) and plenty of natural nitrogen and trace minerals. Water is very important:too much water will drown and not enough will kill off the bacteria. There are many different types of bacteria present in a compost heap depending on the time from the start of the pile.

Different types of bacterial composition will occur through out the composting process. The time of year also determines the rate at which the composting process occurs.

THE INVISIBLE GARDENER'S ORGANIC COMPOST FORMULA.

The best compost formula is one which Mother Nature has already worked out for us.

From the Ocean

From the ocean we take seaweed, kelp, greensand, crabmeal, oyster shell,(anything that comes from the ocean is rich in trace minerals). One just has to send samples to a lab once a year and make certain of no heavy metals or toxins.

PERCENTAGES OF Ocean	N	P	K
Shrimp parts (dried)	7.8	10.0	3.4
Dried ground fish(dried)	8	4.0	2.0
Lobster waste (dried)	2.9	1-5	2-3
Crabmeal(dried)	10.0	2-4	1-3
Crab Meal	2.0	1.0	4.0
Fish MeaL	5-7	5-7	2-4
Seaweed MeaL	05-2	1-5	5-8
Liquid Seaweed	1-2	1	2-5

NATURAL PEST CONTROL

Animal

Animal byproducts	N	P	K
Dried blood	1O-14	1-5	1-4
Feathers	10-15	2-5	05-2
Feather Meal	10-13	2-5	0.5
Eggshells	1.O	.5	.5
Wool wastes	3-6	1-3	05-2
Leather wastes	2-4	1-2	1-4
Bonemeal	0-2	5-12	0-2
Animal Urine.	10-15	1-4	1-2

Dried manures

	N	P	K
Llama	2-5	2-5	1-4
Rabbit	1-3	1-4	05-2
Chicken	2-4	0.8	0.9
Sheep	0.7	0.3	0.9
Steer	0.7	0.3	0.4
Horse.	0.7	0.3	0.6
Duck	0.6	1.4	0.5
Cow	0.6	0.2	0.5
Goat.	1.2	1.3	0.7
EarthWorm Castings	1.0	6.0	1.2
Bat Guano	2.0	8.0	0.5

Plant

	N	P	K
Alfalfa Meal	3-5	1-2	1-2
Cottonseed Meal	6-7	2-4	2-8
Cocoa Bean Hulls	3-5	2-4	2-4
Coffee Grinds	4.0	2.2	5-8
Coffee Hulls	2.0	4.0	4.0
Tea left overs (bag and all)	NA	1-2	1-4

The Mineral Kingdom

	N	P	K
Granite dust.........	NA	2.0	3.0-5.5
Greensand............	NA	2.0	7.0
Gypsum...............	NA	1.0	6.0
Rock Dust.............	NA	2.0	3-7 *

*see rock dust chapter

Greensand, rock dust, oyster shell, rock, phosphate, limestone, granite dust, and dolomite all provide many sources of trace minerals needed by the plants for healthy living.Added to your compost this will increase mineral etc..

Go to your local fish market, there you can get shells which you can dry and smash into a powder using a hammer and an old burlap bag.

The Plant Kingdom

From the plant kingdom you get humus. This comes from leaves, grass clippings, Here are some sources from the plant kingdom: hay, leaves,apple skins(ash), banana skins(ash), coffee grounds, cottonseed meal, peanut shells, pine needles,

A word about Human Hair

Human hair contains nitrogen and other trace minerals. Because of the amounts of chemicals used on hair these days at the beauty parlor the hair will not properly compost. Therefore use hair from a mens salon instead of a woman's, as most men do not use these chemicals. Hair also takes too long to compost within 60 days and therefore should be used in the sheet method described later.

The Animal Kingdom

Obtain any or all you can of: Cattle, chickens, horses, sheeps, llamas, goats, rabbits, and ducks. Best manure is aged at least 6 months.

Trace minerals also come from various sources such as:Oyster shells, coffee wastes, silica sand, bay crab meal, kelp meal,(hair is a good source of trace mineral but should be composted well before using) (hoof and horn meal, bonemeal, and blood meal are all optional). Animal manures are also rich in various trace minerals(always vary your manure source).

To Bin or not to Bin?

When you begin to take into consideration what you have available to work , you will then have to decide whether to go the bin route or the pile route.

NATURAL PEST CONTROL

In short if you need a lot of compost then the piles are easier and faster then using bins. Bins on the other hand work fine for a person with a small garden/home. Three bins min. are suggested. They should be at least 3ft by 5ft by 4ft high and made of wood. There are many books available on making compost bins.

Is it rotting or not?

Can you tell the difference between compost that is rotting or composting? If you can't then your in deep compost!

"If it doesn't smell bad and looks good and dark like rich earth then its done!"

HINTS.

Height, Heat and Air:
Heat is important in compost making. When making bins or piles, the bigger the pile the more heat it will be capable of producing.

Too small a pile, and too much heat is lost. Bacteria love heat and work best within a range of 140-180 F. Too big a pile will compress the material too tightly and make decomposition a slow process. Also too much heat will kill off bacteria. 4 to 6 ft tall is the proper height for a compost pile. This pile will shrink as composting takes place.

This method can also be done during the winter time as the pile is big enough to generate its own heat.You monitor the temp to make sure the pile heats up enough.

About Nitrogen

A proper Carbon-Nitrogen ratio is important for proper composting. Nitrogen is needed to help heat up the compost. Carbon is needed as the fuel. Adding natural sources of Nitrogen (in the form of alfalfa meal, manure or any organically nitrogen rich material) will increase its energy level. Allow for great activity of the bacteria present. Moisture is important in the composting process, but it must not be soggy. I would not concern myself with this ratio just work with what you have keeping in mind the balance re-

quired to make the compost pile work right. The ratio will come out just right if you apply a little bit from each kingdoms as mentioned above.

METHODS OF COMPOSTING.

Indore method explained...........
30-60 day method average pile size...6' wide x 3-5'ft high x 30' long When starting place 6 inch layer from plant waste or animal manure.

He is an example of a schedule to follow, change to suit your needs:

1st Day.
First in center place a large pile of animal manure such as horse or cow or rabbit or Llama pellets. The size should be about 2 feet high. Water lightly.

Place the following layers

Layer in plant waste(hay, grass clippings,leaves etc).

Use your own grass clippings, and leaves to make sure toxins are low or ask if using someones elese's grass clippings).Sprinkle a thin layer of either alfalfa meal, nature meal or composted manure(do not use cat or dog manure) or any nitrogen rich material.

Spray this with a liquid seaweed mixture. You can use Superseaweed(tm) for this as it makes a good compost starter.

Add 1 cup Nitron A-35.

Add a 2 in layer of rock dust covering with a 2in layer of top soil. Water to moisten. Superseaweed(tm) is available at: The Invisible Gardener Inc. Other Compost Starters are available at the following outlets: A-35 Nitron Industries, Inc; they make various compost starters and compost bins. Nitron A35, Necessary

NATURAL PEST CONTROL

Trading Company; a great source of compost starters and soil thermometer, Aerators and many other organic products, Peaceful Valley Farm Supply; another good source of compost starters, thermometers. and various organic materials, Gardens Alive!; they carry many composting products, Mellingers, Inc; they carry a big line of compost starters, organic materials etc. See also Arbico for composting supplies.

When you write to these folks, tell them the Invisible Gardener sent you!.

2nd Day

1in layer plant waste or kitchen wastes,
1in layer alfalfa meal, water to moisten,
1in layer kelpmeal or dried seaweed,
2in layer manure(rabbit,goat,Llama pellets),
2in layer plant waste(leaves, pine needles),
1in layer minerals (rock dust or granite, gypsum, or greensand,oyster shell),
2in layer manure cover,
1in layer top soil, water to moisten,
1in layer rock dust(a good fertilizer and anti fly dust),
1in layer kelpmeal,
Spray with a liquid seaweed such as Superseaweed(tm). cover with a clear plastic or tarp.

3rd day:
Allow to sit
4th day:
allow to sit
take temp(with soil thermometer), leave pile covered
add a 1/2 in layer of rock dust covering
1in layer top soil, water to moisten.

5th Day:
allow to sit

6th day:
take temp, temp should be rising every day till reaches 160 F. The temp may go higher but should stop no higher then 180 degrees F. If the temp does cont to raise add water to cool down.

7th day:
allow to sit
check temp

8th day:
uncover pile and allow to sit open for a day
check temp
check for odors(should smell of ammonia)

9th Day:
turn pile over
spraying with the liquid seaweed as you go.
turning the pile:
This is done by one of several methods...the easiest being using a front end loader to turn the pile over; or

use the shovel technique. The main idea is to turn the pile over in such a way that you are turning the layers together except for the center of the pile. The easiest way to do this is the volcano method.

The Volcano method.

This is when you dig into the center of the pile making it look like a volcano has just erupted, then you throw everything back into the center of the pile mixing it up and watering lightly as you go.

Cover the pile with clear plastic. The clear plastic has several functions:

1: it solarizes the compost destroying harmful bacteria

2: solarization breaks down various in organic compounds and allows for its disposal.

NATURAL PEST CONTROL

3: the clear plastic is a good control against flies laying eggs in your compost pile.

The plastic also protects against rain.

Day 10-15:
Allow to sit
Check temp every two days, if not hot enough add nitrogen rich materials such as bloodmeal, if too hot add water to cool down then add more humus or soil to increase nitrogen:carbon ratio.

Day 16:
Uncover pile
check temp
turn pile as described before .
1lb alfalfa meal
1lb cottonseedmeal use only organic
5lbs rock dust
5lbs kelpmeal
1/2 lb epson salts
1 lb greensand
Stir together in bucket and sprinkle a thin layer as you turn over pile. Then spray liquid seaweed. The pile should be turned over completely, making sure pile is not too wet (soaky) and not dry but moist. Cover pile with plastic if possible.

Day 17-21.
Allow to sit,checking temp every three days. Keep notes on the temp and compare. Is it still rising or is it falling? The temp should continue to raise until it reaches 160 or higher. Then the temp will level off. At the point where the temp starts to drop that is the time to turn the pile over again. This usually happens around the 20-30 day,depending on the weather.

Day 22nd
Uncover pile and turn over while spraying with the liquid seaweed. After this the plastic cover is not needed.

Day 23-30
Allow to sit watering lightly to keep moist. Make sure to spray with a liquid seaweed mixture.

Day 31:
Turn pile over lightly,(aerate the pile by turning over the top layers).

Day 32-45:
Allow to sit checking the temp every 4 days.

Day 46:
Turn pile over while spraying with liquid seaweed mixture

Day 47-60:
Allow pile to sit , water once every three days. check temp every time before watering.

Day 61:
Your compost is ready!!

Working the bin method: The 14 day method

Day 1:
The bottom layer is shredded manure,
The next layer is of plant waste shredded,
The next layer is of rock dust,
The next layer is of alfalfa meall,
Kitchen food wastes ok here ,
The next layer is of old shredded manure such as horse,
The next layer is of top soil,
The next layer is of kelp meal or seaweed meal,
The next layer is of rock dust,
Then add of either rabbit,or goat(shredded) or earthworm castings,
Then add either grass clippings, leaves or both mixed and shredded,

NATURAL PEST CONTROL

Then add a layer either of old horse or old sheep or old cow manure,
Then add a layer of dia-earth(for fly control),
Top it off with a 4 in layer of top soil.

In between each layer you should add the following: mix together equal parts: Cottonseedmeal, Alfalfa meal, Kelpmeal, Epsom Salts, Rock dust or Greensand or SoftRock Phosphate, and old Compost..

The idea is to lightly sprinkle a thin layer of this formula in between each layer, while also spraying (lightly) a liquid seaweed mixture such as a Superseaweed(tm) and Nitron(tm) and Shure Crop mixture(1 cup each added to 1 gallon water(20 drops of superseaweed). Use filtered water.

Your compost bin should have a lid and be made of wood to allow it to breathe.

Day 2:
Allow to sit

Day 3:
Check the temp. write this down in your log book

Day 4:
Turn contents of bin #1 into bin #2(you should have three bins for this purpose), turning over well and adding a sprinkling of the formula as you go and also spraying with the liquid seaweed.
Start the process over again in bin #1.

Day 5:
Check the temp in both bins , note this in your compost log
Turn over bin #2, making sure everything is turned over well while adding a thin sprinkling of the formula.Check to make sure both bins are not too wet or too dry. Add water as needed by spraying(this water is the same liquid seaweed I've been talking about).

Day 6:
Allow both bins to sit

Day 7:

This is your first week!
Isn't compost making fun!!!
Today you turn over both compost bins, checking for moisture and heat.

Day 8:
Then the next day dump the contents of bin #2 into bin #3, turn bin #1 into bin #2 and start all over again with bin #1.

Day 9:
Allow everyone to sit and rest.

Day 10:
Turn over bin #3(sprinkle with formula mixture). Turn over bin #2(sprinkle with formula mixture). Check all bins for water, temp

Day 11:
Allow everyone to rest today, checking for too much water etc.

Day 12:
Check on bin #3, how does the compost look? are there any parts which don't look decomposed? remove that and place in bin #1.
Turn over bin #3 ,water as needed with liquid seaweed and add a sprinkling of the formula.
Check bin#2 , how are we doing here?
Turn over and sprinkle formula and add water(liquid seaweed)as needed.

Day 13:
Two more days to go...
Allow all bins to sit and rest today

Day 14:
Well here it is the day you've been working for.
Check bin #3 and tadum!!! rich compost!!!!

Hints on the bin method

The first thing you must remember is never to add

NATURAL PEST CONTROL

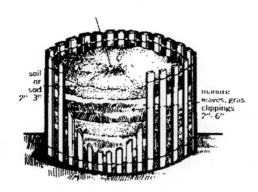

soil or sod 2" - 3"

manure, leaves, grass clippings 2" - 6"

raw waste into bins #2, #3. These bins are for finishing the composting process. Bin #1 is where you may also add kitchen waste such as rice left overs, salad left overs etc. Never add meat or cooking oils as this will stop the composting process and you may even have to dispose of the compost. You may add egg shells, coffee grinds,to the first bin only. Never use animal manure that comes from meat eaters such as dogs, cats, humans, lions, tigers, etc.

The zoo is a good place to get various animal manures. Rock Dust is excellent for helping the composting process.
See Resources Directory for addresses on companies selling organic supplies mention in this book. Please support IGA members.

More Reading:
There are many good books available on compost and composting! See "The Rodale Guide to Composting", "The Encyclopedia of Organic Gardening" by Rodale see also "Secrets of the soil", by Tompkins and Bird.

Composting Problems

Compost Too Wet

Problem: compost is either too wet(soggy) or stays slimy.

Solution: poor aeration, over watering.

Your compost needs to be turned over more often,allowing air to get at the center of the compost which will increase the energy levels of the compost and if enough nitrogen is available it will heat up. This will also help to dry it up some.

You are adding too much water when you are watering in the compost. Sprinkle only a light amount of water in between layers. Use filtered water for best results as chlorine water kills beneficial bacteria. Do not allow too much rain to fall on it. Cover with a tarp.

Not Cooking up

Compost needs energy to cook up. bacteria need energy to work. What they need is minerals, carbon , nitrogen rich materials. The addition of Rock dust to your compost setup will increase minerals available to bacteria. The addition of animal manures is very important to adding bacteria, minerals and nitrogen to the compost pile.

Compost Too Dry

Problem: My compost is too dry !

Solution: Water it! Best water is filtered water. Water in between layers. if pile dries out then cover with a trap. Be careful not to over water. without water the pile will not cook up,a good idea is to use a sprinkler can that the water has been allowed to sit over night in. Add a little seaweed or fish emulsion to increase bacterial activity. Nitron A-35 is good for this too. Nitron's number is 1-800-834-0123 tell'em The Invisible Gardener sent ya! The Garden Filter is available @ 1-800-354-9296

Compost has Bugs in it!

The four most common bugs you will find in your compost are pill bugs, sow bugs, ants and earwigs. All of which are highly beneficial to the compost system. However they have their place and time. When you have any bugs in your compost then either the compost is ready or you have to make the compost get hotter. The heat will also drive the bugs away while composting!

I suggest that you turn the pile over while adding high energy sources to the compost. Such soures as rock dust, animal manure and alfalfa meal. If your

NATURAL PEST CONTROL

compost is ready to use and you have bugs in your compost then I suggest that you set the compost out in the sun for a few days. This will remove the bugs from your compost.

Compost Attracts Animals

Actually the compost itself does not attract animals but rather the animals are attracted to the food that you are throwing into the compost.

There are several ways to avoid this problem:

Use a 3 bin system. Only throw food into the first bin. Make sure its layered with other organic matter and combined with a high energy source such as rock dust to insure heating. Heat will keep an animal out. The bin should have a lid which is kept closed restricting access to animals.

You can pre-compost your kitchen food before you put it into your compost bin. This is done by using a small waste bucket which is filled half way with compost/soil. place kitchen scraps into this bucket and mix in with the soil. Dump into compost bin once per week replacing with new compost/soil.

You can run your kitchen waste through a blender and then pour into compost pile.

Get a cat! just kidding

Compost too Hard to turn

There are many reasons why your compost is too hard to turn, most of them are physical reasons, either too old or unable to turn.

I would suggest that you try one of the following:

Hire someone to turn it for you
Get a tumbler that you can hook up to your bike
Get into vermiculture(let the worms eat it for you).

If your compost is really very hard to turn then I would add more lighter materials to the pile such as leaves, etc.

Compost too Hot

If your compost is getting too hot then you are adding too much nitrogen rich materials such as chicken or animal manures. I would suggest that you re turn the pile while adding carbon materials and sprinkling a little water to cool it off. The addition of soil would be good here also.

Compost Smells Bad!

If your compost smells like ammonia, then you are adding too much nitrogen rich materials(like raw manure which is rich in urine). You may also be adding too much water. If it smells rotten and there are a lot of flies around it, you are adding too much kitchen wastes without properly turning it. Make sure that you are adding enough of the various materials needed by the compost pile; green wastes such as grass clippings,brown wastes such as dried leaves, kitchen wastes, minerals such as rock dust, bone meal, etc, bacteria rich such as animal manures. Turn your pile over more often!

It's an Art

Remember that making compost is truly an art. The canvas of this artist is everywhere you look. The earth is one big compost pile and the art form forever changing and the artist forever changing too.!

Making Compost from your kitchen wastes.

The art of making good compost from your kitchen waste is one that you learn over a long period of time. Experience is the best teacher here. What we will do is set you upon a path that will lead you in the right direction, pointing out the pit falls along the way.

Obtain a 5 gallon wood bucket. Wood is best . The pvc which most plastics are made of is not good in the ground because of its possible health hazards(we will be doing a report on this subject in our next newsletter). The kind that dog food comes in is ok; for temp use, equally good is a painters metal bucket (unpainted). Cut off the bottom and bury in garden. Bury at level with the ground, so that the top will be level, or slightly above the ground level. Make sure you have a top for it as it will need to be covered. You will need at least two of these buckets, one at either end of your garden.

Set up several plastic bins for recycling paper, cans, and bottles and re-using glass.

Choose a composting area for producing your main source of compost. This area can be done in several ways. One way is to make wooden bins, always use untreated and un painted wood. A good compost bin made of redwood, oak or pine, should last you at least years, This can be extended depending upon location, exposure to the

weather and construction. Bins can also be made of rocks, concrete blocks, logs, crates, pallets, wire screening, or just about any thing you can think of that will hold the compost in while allowing the air to circulate(in our next newsletter we will show you how to make your own compost bins).

Now that you have this simple set up down, you can begin to save your kitchen wastes as well as green waste from the yard. The kitchen wastes should not contain bones, meat scraps or lard. Place kitchen wastes into the buckets in the garden along with some compost from your compost bins, as well as a little bit of soil. You can now also begin to put your kitchen waste into your compost bins. You should have three bins, ea 3 x 3 ft. (see example above).

In bin number one you place your kitchen waste. This is the only bin that raw materials go into. Do not put anything raw into bins two and three. Back to bin number one. Place a layer of grass clippings ,(no more then 2 inches), then a layer of kitchen wastes, (no more then 2 inches), then a layer of soil(about 4 inches) then a layer of rotted ma-

nure, rabbit , chicken, steer, horse, or ?. A good source for this stuff is out in the country. Somethings are available in your local nursery. Look for horse farms or egg farms in the yellow pages. If you know of some one who makes compost and would sell you a small amount then you can use this to jump start your bins. The bacteria would really help your compost along. Purchase a good natural organic fertilizer(see Resource list

Sprinkle a little bit of this stuff in between layers for best results. Try soft rock phosphate which is an excellent source of iron, magnesium, calcium, etc.

When bin one is filled up in this way throw it into bin two and start again on bin one., When this bin is again filled up by the layering method then take whats in bin two and toss that into bin three, and whats in bin one into bin two and start fresh with bin one.

It should take about 1 month to completly fill one bin.What comes out of bin three is ready for use.

COMPOST PRODUCTION

Going to the library, I found no less then 15 books on the subject of compost production! It is no wonder that many people are confused about this subject. Thruout the many years in which I have written about compost, I have tried to demystify this subject.

In my classes, workshops, and lectures, I have make the connection between Natural Pest Control and Compost. Simply put you cannot expect to control pests naturally if you do not use compost as your main source of nutrition for your soil and plants. On page 5 in this newsletter is a chart which shows the relationship between nutritional levels, stress levels and pest levels.

If made correctly compost is a complete balanced food for the soil and for the plants which it supports. Many people believe that a pile of leaves is compost. When I make a housecall and examine their composting process, I find that most people make their compost strictly from the materials at hand, such as leaves, grass clippings and sometimes kitchen wastes. What is wrong with this picture?

The problem with this picture is that unless your property has rich virgin soil, your compost will be lacking in many minerals, bacteria etc., essential for proper plant growth. If any one mineral, bacteria is missing stress will result. Controlling stress is the single most important factor in natural pest control.

To properly make compost, you must constantly introduce these minerals, bacteria, enzymes into your compost process. Let me give you an example: If you ate only food which you grew on your property, and you did not constantly introduce these minerals, bacteria, enzymes into your compost, you would not be healthy for long since your body requires a constant source of minerals and a variety of food sources in order to stay healthy. Got it?

It is the same with the compost, the plants, the soil and everything that depends on it for its health. By not providing the soil with a constant source of minerals, bacteria, enzymes, you are increasing the stress of the system. A stressed system will attract pests, diseases and will eventually not support life.

Chemical fertilizers do not support the natural system. They provide only a small amount of minerals(if any) and promote the sterilizaton of the soil. They kill off the earthworm population and destroy bacteria and the natural enzymes. When you use chemical fertilizers you are playing the numbers game.

Have you ever noticed that when you buy a bag of a chemical fertilzer that there is three numbers printed on the front? Something like 12-12-12 or 30-10-10. The actual numbers will vary according to the amounts of Nitrogen, Phosphorus and Potassium available in that fertilizer. Some fertilizers will have four numbers with the fourth number being Sulphur. This is all that the plants will get period!   Plants need over 76 minerals for survival and good health!

To make a long story short here is a simple formula to follow when making your own compost:

Take something from the ocean (like seaweed, greensand, Superseaweed,etc.), take something from the earth (like rock dust, soft rock phosphate, etc), take something from the plants (like leaves, grass clippings, kitchen wastes, etc.), take something from animals (like animal manure, bone meal, blood meal, etc), take something from natural bacteria (like Nitron A-35, Shure Crop, Superseaweed, etc.). Don't forget when you make compost to sprinkle filtered water inbetween the layers (Nitron sells a good garden filter for this also).

Layering is a good method of insuring that you have the proper carbon\nitrogen ratio required for it to heat up. This heat is the result of the bacteria throwing a party and eating it all up. Air is also required by the compost in order to continue this composting process.

Finally Compost production must be a fun process for you and your family. Involve everyone in the turning process, and in its application.

Happy Growing, Organically
of course!
Andrew Lopez

Everything you need to design and implement your municipal home composting program

Keep It Off The Curb

Your Complete Step-By-Step Manual For Home Compost Program Management

By Harmonious Technologies and top composting experts across North America

Keep It Off The Curb is a practical desktop manual designed specifically for use by recycling coordinators and extension agents.

- Save thousands of staff hours

- Benefit from the experience of hundreds of successful programs, including 30 case studies

- Utilize a ready-to-copy brochure, survey forms, workshop signs and more, plus a comprehensive list of additional educational resources

- Avoid pitfalls and slowdowns

- Learn how to sign up 25% of your residents in one weekend

Call 805-646-8030

Harmonious Technologies
The Home Composting Specialists

THE WORLD IS OUR GARDEN.

The Invisible Gardeners
of America
SINCE 1972

An Educational Organization dedicated towards better living thru Natural Alternatives to Chemicals used in Pest Control.

NATURAL PEST CONTROL

Chapter # 16
Rock Dust

The Elixir of the Earth

To grow a greener lawn, have healthier trees, and cultivate bigger vegetables, the soil needs to be enriched. Before reaching for that bag of nitrates or other chemical fertilizers, the conscious homeowner or farmer should stop for a moment to consider what needs to be put back in the soil to enhance its life-giving properties.

Like the magnified form of the human body, the earth has the wondrous capability of healing itself. When forested areas of the world use up the nutrients in the soil, the earth has a built-in remineralization system that can be learned from and applied right in the back yard.

The process is known as remineralization through rock dust application. John Hammaker, a research scientist in Massachusetts, postulates that each ice age in the history of the earth regenerated its topsoil.

When the planet's forests deplete the soils of nutrients, they begin to die off, releasing their stored carbon into the atmosphere. All this carbon builds up in the atmosphere, creating a "greenhouse effect" and causing a rise in the earth's temperatures.

This heat is most concentrated around the equator where the sunlight is greatest.The higher temperatures cause evaporation, and the moisture rises, creating a vacuum underneath which pulls in cooler air from the polar regions. As the air is pulled in from the North and the South, another vacuum ensues, pulling the warm moist air towards the poles. During this time there is a lot of strong activity like typhoons and hurricanes that occurs in the earth's sub-tropical regions.

When the moist warm air from the tropics arrives in the polar regions, it hits the cooler temperatures and condensates as snow. The snow builds up, and the weight causes it to pack into the ice and push southward from the north pole and northwards from Antarctica. At this time there is an increase in earthquakes and volcanoes caused by the extra weight on the continents.

The glaciers descend, grinding up all rocks and mountains in their path, remineralizing the soil. When the forests again take root, they absorb the carbon from the atmosphere through photosynthesis, and the ice age diminishes.

Hammaker has extended this hypothesis to explain what is currently occurring on earth right now. According to Hammaker, earth's inhabitants have accelerated the onset of the next ice age through the burning of fossil fuels and the deforestation of large forested areas like the Amazon.

Hammaker says the only way to stop the glaciers from knocking down our back doors within this generation or next, is to reduce our dependence on fossil fuels, reforest the cities and the country side. by replanting and remineralizing the earth through rock dust applications.

While Hammaker advocates going out and remineralizing all forests and fields, the average homeowner usually does not have the time to take on such a large task. However, simply by taking care of one's yard and garden through natural means, the accumulative effects will yield significant results on a global scale.

NATURAL PEST CONTROL

About Rock Dust

The amount of minerals and the quality rock used to produce rock dust depends on the location of the Quarry, and the mining process. Many companies sell rock dust, but it is best to inquire about what minerals it contains and if there are any chemicals added. If their manufacturing plant is nearby, ask to take a tour. Some companies simply sell the dust that is left over from manufacturing other products. This may not have the necessary minerals for plants and trees, so it is best to buy rock dust from those companies which make it specifically for this use. They will be happy to provide a lab report for you.

Rock dust is usually high in calcium, iron, magnesium, sulfur, and more than 100 other trace minerals. The PH level is often very high (eight or nine) and therefore must be used in small amounts, combined with compost, with peat moss if acid soil is required, or made into a liquid form. The rock dust does not dissolve when mixed with water, but forms a colloid, making it instantly available to the plants it is sprayed on.

Using rock dust can replace many other natural mineral products which are harder to get and more expensive.

The Bacteria Eat First

Not all the food that you add to your soil makes it to the plants. Initially, they must be broken down by soil microbes. The microbes live in many different areas of the soil, some live on the root hairs of many, some live only deep in the ground. The bacteria which live on the root hairs of plants convert minerals found in the soil, into a different form of the same mineral which is available to the plants. Microbes tear apart or combine minerals. Also when the microbes die they leave behind minerals in a changed form which is also available to the plants as food.

Mix The Dust

For best results mix different sources of rock dust together to get more complete trace minerals. There are many companies now offering rock dust. We have listed as many of them as possible in our Resource Directory.

Making your own Rock Dust

While this is not the easiest way, it can be done. Obtain a fifty gallon steel drum, weld bicycle gears at both ends. Hook up a bicycle through the chain gears. Set up the front wheel on a non-moveable base. Get a large round river stone to place inside. Make an opening which can be closed and locked. Then place local rocks into drum and exercise while crushing the rocks! Most soft rocks will work well. You can also crush lobster tails, clams, and other seafood materials. Add oyster shells to increase calcium.

NATURAL PEST CONTROL

How to use Rock Dust

For Compost Production

Using rock dust in compost production increases its energy level by adding minerals, and increasing the activity of bacteria. The increased bacteria stimulates the composting process. Rock dust also helps heat up the compost pile.

Add a thin layer of rock dust to the compost pile, alternating between layers of grass clippings, manure, kitchen wastes, etc., Water as needed. As rock dust is high in PH level, use small amounts and add something acidic like leaves or pine needles. Never add chemicals to the compost. To lower PH level if it is too high, add one quart vinegar with five gallons water and sprinkle over pile once per week.See chapter on compost for more information.

For Pest Control

Rock dust is also very effective in controlling pests such as snails, because of its high silica content(67%). Dust lightly around the garden, allow to sit for 24 hours , then water well.

Trees

Trees also love rock dust. Always use small amounts as the PH level can irritate high acid based trees. A large tree should be given four coffee can full of rock dust spread evenly beneath its canopy, starting two feet from the trunk to ten feet past the furthest reach of its branches. Sprinkle evenly then spray down with hose. Use a garden filter to filter out the chlorine from the water.

In The Garden

Dust around the vegetable plants, allow to sit overnight, then spray down with filtered water. Mix into a liquid form and add seaweed. Spray this on the plants leaves, this increases the energy level of the plants, assisting them in fighting off pests. Use peat moss or composted aged wood to maintain a balanced PH level. Your vegetables will have increased mineral and vitamin content and will be more nutritious.

For the Lawn

When used on lawns , rock dust will increase color and encourage deeper root systems. Using rock dust will decrease your need to add any high nitrogen . Rock dust will also increase the effectiveness of compost applications. Add 50 pounds of rock dust per 1,000 square feet of lawn. Apply twice yearly.

After dusting lawn, water well. Take care not to breathe in the rock dust when you are dusting your garden, lawn, trees, etc.

Some Benefits of Rock Dust

Plants achieve physical completeness more quickly.
Due to its high mineral content, rock dust helps in all aspects of the plants biological process.
Plants are healthier.
Plants are insect free.Less prone to attack.
Plants can handle stress better.
Plants live longer healthier lives.
Rock dust is a natural fertilizer.
Non polluting, biodegradable.
Rock dust is inexpensive compared to other mineral products.
Rock Dust increases mineral content of the soil and

NATURAL PEST CONTROL

the plants growing on it.

Shopping for Rock Dust

Some things to look for in a good quality rock dust. You'll want a very fine dust, apprx. 200 mesh. This is important because the finer the dust the more rapidly it is available to the plants and the microbes which have to eat it.

The location it is mined in is important as it gives us a clue as to the forces that went into making it. High energy always makes high energy. Ask for a lab report. What PH is it?

Don't use cement for construction use. Tell them what you want to do with it. Don't use any rock dust that has any type of chemical added to it.

The Test

A good test is to fill a clear glass half full with your sample and cover it with 3 inches of water. Shake it up and allow it to settle. The dust, silt and sand will settle into three different layers, with the dust settling on top. This will give you the percentage of how much of each you have. The finer the grind the easier the bacteria can get at it. However I have also found it to be true that small chunks, less then 1/4 inch, are good for the soil as other creatures will eat it also. Another thing is that these chunks will break apart later providing additional food.

Another Test

Take any magnet and place on rock dust. If it clings to the magnet it is of the right energy polarity.

Making Rock Dust Milk

Slowly add water to a cup of rock dust. Stir slowly until dissolved. Add to 1 gallon water, allow to sit, then strain into sprayer. This makes the food instantly available to plants sprayed.

Types of Rock Dust

Rado Rock comes from Canada straight from the glaciers. Planters II comes from the Colorado Rockies. Earth Wealth comes from the San Gabriel Mountains in southern California. Azomite comes from Utah. Rock Phosphate is a well known rock dust, excellent when finely ground. New Jersey Greensand is also a very nice rock dust but can be expensive to buy depending on which side of the USA you live on.

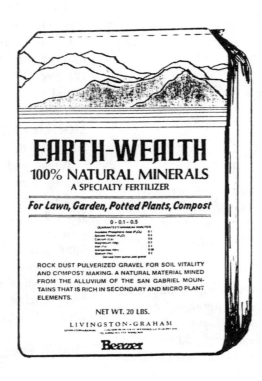

EARTH-WEALTH
100% NATURAL MINERALS
A SPECIALTY FERTILIZER

For Lawn, Garden, Potted Plants, Compost

0 - 0.1 - 0.5
GUARANTEED MINIMUM ANALYSIS

ROCK DUST PULVERIZED GRAVEL FOR SOIL VITALITY AND COMPOST MAKING. A NATURAL MATERIAL MINED FROM THE ALLUVIUM OF THE SAN GABRIEL MOUNTAINS THAT IS RICH IN SECONDARY AND MICRO PLANT ELEMENTS.

NET WT. 20 LBS.

LIVINGSTON-GRAHAM

Beazer

Some Sources of Rock Dust

Contact your local state farming organization or local organic grower for local sources of rock dust.

Resources

For complete addresses see Resource Directory. Peak Minerals-Azomite, Bio-Plus Glacerial Rock Powder, Planters II Trace mineral fertilizer, Rado

NATURAL PEST CONTROL

Rock Glacial Dust, Glacerial Mineral Dust, Earth Wealth Rock Dust, Peaceful Valley Farm Supplies, Nitron Industries

Other Sources of Trace Minerals

ARBICO,Neccessary Trading Co, Ohio Earth Food, Brickers Organic Farms, Garden-Ville, Orol Ledden and Sons, Peaceful Valley Farm Supplies, Nitron Industries. National Research and Chemical company(has a large Organic dept too).

A Good Organic Soil Testing Lab is Timberleaf Farm located in Albany, OH see Resource Directory for address and phone number.

ROCK DUST ROCK DUST MORE ROCK DUST

"One persons rock dust is some else's gravel, pond sand, quarry silt, bug dust, swamp sand, crusher dust, fill sand, or ? Always tell them what you plan to do with it."

When I tell people that I have rocks in my compost they laugh and say "Sure you do!", and I say "Yes I do and I also have rocks in my garden"! Rock Dust that is. Up until recently I was not able to find a local source of rock dust for my compost or garden. But with the help of "Secrets of the Soil" and other such books , more and more companies are making this great stuff available to us organic growers.

In the may 92 issue of Organic Gardening Magazine, there is an excellent article on Rock dust which will help you to understand why its so good and it also gives some sources. As I mentioned in my last newsletter, rocks are formed inmany ways, the key to getting good rock dust is how it was formed in the first place. The more active the formation process the more "alive" the rocks formed and the more "energy" will the rock dust have.

To me, energy means several things but in this case it means how much minerals it has and the quality of the minerals it has. The higher the energy, the higher the quality of the minerals available. With rock dust quantity is not as important as quality. The key to rock dust is the quality of the minerals available.

IGA Club RESOURCES

Support these IGA members

Earth-Wealth Rock Dust
14941 Daffodil,
Canyon Country, CA
91551
Ask for Walt Zmed
(805) 298-0791

For more information:
Acres USA
10008 E 60th Terrace,
Kansas City, Missouri 64133
(816) 737-0064

Down to Earth
800 Park Way
Lake Elsinore, CA
92530
909-674-8558
Ask for Danny

For Trace Minerals:
Nitron Industries
4605 Johnson Rd
PO Box 1447,
Fayetteville AR 72702
(800) 835-0123

During the course of this past year , I have begun to experiement with different sources of rock dust, mainly the local sources as shipping them from afar is quite expensive. There are several interesting things which I have found about rock dust in general.

#1 The PH level is quite high. This corresponds with the fact that most animal manures are also quite high in the PH levels. This is not bad and not good as too high a PH will bond up your nutrition but the important thing here is that certain bacteria are only found in the higher PH levels. The PH will drop once these bacteria are finished and have died. These bacteria are important to the soil and its transferance of nutrition to the plants.

Once the PH levels begin to drop, different bacteria come into play. In this book I have discussed the different ways Rock dust can be used in the garden, for trees, on the lawn, etc.

For more information on rock dust contact **Remineralize the Earth , 152 South Street, Northampton, MA 01060 U.S.A**
Tell'em The Invisible Gardener sent ya!

Chapter 17
Growing Your Own

THE VEGETABLE GARDEN is a very important link between us and the earth. As our garden grows we grow. When we grow our own food we know exactly what has been used to grow this food. We know that the fruit, the vegetables, the herbs that we grow are rich and full of nutrients necessary for our health. Our backyards can provide us with good homegrown food. Growing your own food is a very important aspect of maintaining a healthy mind and body. Even more important is growing your own food organically without the use of chemicals that pollute ourselves and our environment. If our home grown food is to have any nutritional value it must be grown in healthy live soil. This becomes even more important when you realize that to succeed growing organically, good healthy soil is necessary.

Steps to creating your garden paradise.

Raised Beds or not

The decision to have a raised bed or to grow in the ground depends upon
1...the condition of the soil
2...the amount of space you have to work with
3...the amount of time you have
4...the amount of water you have available to use
5...the amount of money you want to spend

A few points about Raised Beds

1...raised beds allow you to grow from 4 to 7 times more food in the same area then in the same ground.
2...raised beds allow you greater control over watering costs
pests(from gophers to ants)
crops(especially extending harvest periods)
3..raised beds allow you greater control of the soil being used.

Raised Beds are easy to make and you can recycle (re-use) various materials in making the beds. You can use rocks, wood (untreated, unpainted of course), tree logs, bricks. You can use bottles to make the sides of your raised beds. Stick upside down in ground. You can use clay on the inside to mold and hold them together or you can use sand and pack it in or use rope to tie them together. The idea has a lot of potential for re-using glass around the home. Untreated Redwood makes good wood for raised beds. A good size for a raised bed is 10 ft long by 5 ft wide by 12 inches tall and wood should be 2 in thick. This size is big enough to feed a family of 4. Pine, fir and almost any kind of strong wood is all right as long as it is untreated. Never use Rail Road ties that have been treated. This can be very bad for your health as well as being bad for the garden (it kills the bacteria and earthworms in the soil and can end up in your food).

If you decide not to use a raised bed then you must prepare the area first.

Laying out your garden

Draw a plan for what you want to grow in your raised bed. Keep the tallest on one side with the smallest to the opposite side, with consideration to light. If you have more then one raised decide which plants will grow where. Change this design every year so as not to grow the same plants in the same place year after year.

NATURAL PEST CONTROL

Choosing a location for your raised bed is very important. The location needs to be close to the kitchen to provide easy access for the cook. The location must provide at least 6 to 8 hrs direct sun, with the more the better. Must have proper drainage. So take a walk around your place and see if you can pick the perfect spot. Another consideration is water, it must have a close source.

Garden Filter

Never use city water in your raised bed. Many cities have chlorine(or ?) in their city water. Chlorine kills bacteria, that is what it does best. However an organic garden requires natural bacteria in order to function correctly. A garden filter will help keep your garden alive! You will notice an increase in worms, and in the gardens overall health. An excellent Garden Filter is available thru IGA(IGA members only) at 1-800-354-9296. Call for more info.

Putting together the Raised Bed

A good raised bed should be at least 4ft x 10 ft x 12 inches high and 2 inch thick wood. Use non treated wood. The wood can be screwed together for easy break down when needed. If you have gophers in your area, you will want to screen the bottom with extra heavy chicken wire. The size of the bed depends on the area and amount of space you have to use. An ideal situation is to have two or more raised beds. More beds allow you to rotate the beds and allow one bed to grow green cover which can be turned over.

You can always make your beds out of any natural material readily available such as rocks. logs. bottles, etc.

Filling in the Bed

The following should fit just right into the raised bed. You will have to use what you find in your area. Start out with a good layer of old horse manure. To this I add either LLama pellets or Rabbit pellets(nature's time released fertilizer). Add 20 lbs of rock dust or any other trace mineral source. Add 20 lbs bone meal, 20 lbs alfalfa meal and 2 bales aged wood. Mix well. Water well(water slowly to allow soil to absorb). To this mixture add 500 lbs compost(if you have it) other wise add enough old horse manure to fill up to 4 inches from the top. Add another 2 bales aged wood or KRA wood product or any light soil. Add another 20 lbs rock dust, 20 lbs bone meal, 20 lbs alfalfa meal. The aged wood will insure the PH will be at the right place. A good PH for the garden is 6.5 to 6.8.

Blend everything in together, watering as you go. Finally add enough mulch to fill the raised bed up to the top. Remember that this soil will settle after a few days, so keep a few bags of mulch handy to fill to top when needed.

COMPOSTING

The Secret to Growing Organically is Compost. Making good compost is a special art that we all must learn if we are to become Master Organic Gardeners.

GREEN COMPOSTING

30% of our landfill materials come from this area of our wastes. Grass clippings, leaves, etc. make great additions to our composting system. A shredder will help to speed up the composting process.

NATURAL PEST CONTROL

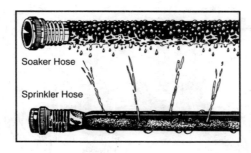

Soaker Hose

Sprinkler Hose

It pays to be able to save all of your kitchen wastes. A small container would be useful to have in the kitchen area for this purpose. Make sure it has a lid. We Compost all of our kitchen wastes and recycled paper wastes as well as the wood from the matches we use, etc., any thing that will compost is saved. Learning to make compost is a great way to take something that is being thrown away and turn it into food for the soil, for the plants and food to us.. This is recycling at its best! Make sure that you empty the container every day. I suggest that you can compost the kitchen wastes by layering into your compost. You can also bury the kitchen wastes into your garden allowing the worms a meal. Adding some rock dust to kitchen waste will help reduce smells and flys, and also increase microbial activity.

MULCH

Mulching is a very important part of the organic garden. It is an excellent way of recycling. It is always best to compost your mulch before using. Never mulch around plants with freshly cut mulch from trees or grass clippings. This will burn the plants.

What is the difference between mulch and compost?

Compost is the food, and mulch protects the food from the elements like rain and the sun which will dry it out. Mulch will hold water and not allow the soil/compost to dry out. A good mulch is made from aged wood. This is wood which has been recycled from cut trees and composted organically. Often companies add urea to their compost believing that the compost needs this "chemical" nitrogen. So ask before you buy! Urea based products are very detrimental to the soils organisms and should not be used. Instead well composted horse manure can be added. Rock dust will also work well here as rock dust will bind

the nutrition together.

The Drip System

A good drip system is important. A soaker hose will work very well here. The soaker hose can be buried about 2 to 4 inches from the top and can be moved as needed. Remember where the hose is to avoid damaging when planting. A battery timer will help to control the water and is easy to operate. Place a garden filter between the hose and the timer. Your garden will love this extra touch.

Buy earthworms for your Garden

Earthworms will love your raised bed. Give them a head start by buying red wigglers which is the best kind for this use.

Planting the Vegetables

Start by making a list of vegetables that you like to eat. If you haven't grown a garden before, I suggest that you take a gardening class to help you out. Consider joining a garden club as they are a great source of not only help, but seeds, resources etc.. The varieties that you choose to grow will be important. Learn which will grow in your area. I also suggest that you get heirloom seeds instead of the commercial ones. Try Seeds of Change, Abundant Life, Native Seeds, Heirloom Garden Seeds, Native American Seed, Native Seeds/Search, Bountiful Gardens, Peace Seeds, Plant Finders of America, Ronniger's Organic Seed Potatoes, Seed Savers Exchange, Southern Exposure Seed Exchange, For more resources try Acres USA. See Resource Directory for phone numbers and address to write to.

Mix Flowers and Herbs

It is a good idea to plant flowers and herbs along with your vegetables for best insect protection. Try 50% flowers and herbs along with your vegetable garden.

NATURAL PEST CONTROL

A raised bed can be protected from extreme heat and cold by placing over it a sheet of plastic nailed down to the wooden sides. This will also help to allow new seedlings to grow and become established.

Feeding

Use only natural organic fertilizers for best results. Chemicals will only destroy the balance of your garden. Allow time for this balance to occur. **See making your own fertilizer chapter.**

CONTAINER GARDENING

Growing in various different containers is a wonderful way to reuse many objects.

Follow a few simple rules

Never use containers that have been used to hold toxic waste materials. Try growing only in natural containers made of wood, clay, rocks, logs, boxes, crates, etc. Try not to use plastic containers. Glass makes good sides for a bed. Always provide adequate drainage. Clean Containers before using.

Tools

Tools are very important to you and your garden. Use the best tools you can get. I often find that it is not important if a tool is new. If the tool does the job that it was meant to do, keet in good shape. Some old time tools are great to use!

Companion Planting

There are many books which cover companion planting. Here are a few:
The Organic Gardener's Home Reference by Tanya Denckla, Environmental Gardening by Karen Arms ., Rodales Books, try Acres USA Bookstore for more.

An Organic Vegetable garden is an essential part of survival and a very important part of helping the earth to heal herself. Thru proper organic growing of our own fruits and vegetables we can drastically reduce chemicals in our environment.

Pollution begins at home. Lets start by not using chemicals in our food production.

It is not true that vegetables grown organically will be imperfect and full of holes. It is not true that by growing organically we will starve and that farmers will not be able to produce the same amounts of food as they do now.

When a farmer says that you have to use such and such a chemical for such and such a problem, what the farmer really means is that he does not know how to do it organically.

So lets begin by covering some of the basics of growing organically. If you follow these basics you should be able to grow organically without the chemicals.

Basic # 1

Do what you do best. If you are inclinced to gardening then it will all come naturally to you. Follow your feelings about what to use or not to use, what to grow or not to grow. Remember experience here is the greatest teacher.

Basic # 2

Nutrition vs Stress

Provide proper nutrition, in the amounts needed at the time needed. Organics are best suited for the purpose. Never use high nitrogen in your garden. This will only cause stress.

#2A Compost is the single most important aspect of growing organically. The better the compost the healthier the soil and the less the stress. Compost properly made will provide your soil/plants with all the nutrition they will need.

Basic # 3

Control stress and you control the pests. The greater the stress, the greater the imbalance. Always seek balance in your life, in your garden.

Basic # 4

The greater the diversity the greater the balance. Allow the insects to control themselves, helping the good guys whenever needed.

Basic # 5

Learn your vegetables. Learn what will grow in your environment. Try different varieties. Learn when to plant, when to harvest, when to rest.

Happy Growing!

We re-use our bottles. They make great raised beds!

In this issue we will be looking at one way bottles can be re-used: by turning them into raised beds, path ways and by using them to control pests as well as to provide homes for many predators!

The first step is to set up a drop off point in or near the kitchen, Here all the bottles used around the home will be deposited after their use. In order to conserve water, I do not suggest that you use water to remove the labels, instead to place the bottles outside (near where they are going) and allow the sun to help you peel off the labels. You can also use a knife to scrape most of the labels off.

Take the bottles and place upside down into the ground along where you want a bed to be. The shape of the bed can be any thing needed to match the surroundings, such as beds with a path around them for walking. The soil should be dug up enough so that the bottles easily push into the ground half way up the bottle.

Caution! Be careful when handling bottles , bottles can break and glass cuts! Do not use a hammer on the bottles! Use instead a rubber mallet and lightly tap the bottles into place.

The corner bottles can be right side up and of a larger variety, usually wide mouthed. This bottle is filled 1/2 way with water in which a tablespoon of wine is added. The cheaper the wine the better! Or if you have fruit trees you can place fruit scrapings into the mixture and the fermentation process will attract pests which will dive into this swimming pool and die. Cider works well too! Also attracts slugs, snails, etc..

You will find that bottles come in many different sizes and that some are better suited for raised beds, then others, such as wine bottles are bigger and make for taller beds while beer bottles make smaller beds and good path markers.

Baby bottles and other wide mouth and smaller in size make good path steps when placed upside down and filled around the bottle with soil.

Bottles have a special effect on the garden as a whole and the beds in particular. The glass bottles are hot during the day and cool at night Also the electrical charge which glass attracts and holds is not liked by snails or by gophers either!

DONT PANIC
ITS ORGANIC

THE INVISIBLE GARDENER'S KITCHEN
"POLLUTION BEGINS AT HOME"

" Sewer Sludge needs to be properly processed which means the removal of all heavy metals, pesticides, etc., Sewer Sludge must be treated as a serious problem to our health, our childrens and to the health of the world"

Safe Green Catalogs

Whats the Problem with Sewer Sludge? Take any sewer sludge product on the market today and send a sample of it to your favorite soil testing lab and ask them to look for heavy metals (and if you have the money , to look for pesticides, etc)., they will find more then you bargained for! Why? Mainly due to our eating practices such as meat eating, seafood, etc. All of which is increasing becoming more polluted. Remember what you are eating is the end result of many meals before you. What ever you have just eaten, ate also and what it ate also eats concentrating each pollutant eaten. The higher up the food chain you eat, the higher the concentration of toxins there will be. The waste of each creature is proportionaly concentrated with toxins, pollutants, pesticides, etc, tjhe higher up the food chain they are. Mankind is currently at the top of this food chain. Bingo, Bingo!. There are new developements in the processing of this wastes thru the use of bacteria, etc but most companys still do it the old way of drying (which furthur concentrates it), screening, cleaning (which they call composting and it is not), they remove glass, metals, etc., remove most of the smell and blend it in with other products which are sold to you the consumer for use in your gardens, lawns, vegetables, etc.. They all should have warning labels warning people of the higher levels of heavy metals in that product. Remember small amounts can add up to larger amounts given enough time.

ABC Organics
Box 5313 Santa Monica CA
90409-5313
310-450-9962

Los Angeles Green Pages
24955 PCH
Suite C-201
Malibu, CA, 90265
310-456-2163

The Sharper Image
650 Davis St
San Francisco, CA
94111
800-344-4444

Gardener's Supply Company
128 Intervale Road
Burlington, VT 05401
800-444-6417

Gardens Alive !
5100 Schenley Place,
Lawrenceburg, IN
47041
1-812-537-8650

Smith and Hawken
Two Arbor Lane,,
Box 6900
Florence, KY,
800-776-3336

Jade Mountain
P.O.Box 4616
Boulder, CO,
80306-4616
800-442-1972

Gardeners Eden
P.O.Box 7307
San Francisco, CA,
94120-7307
800-822-9600

Appendix

"Everything you need to know to develop your own Natural Organic Spraying Service for your Own Community."

In Order to succeed at a Natural Gardening Service, you must work within both mans laws and natures laws. Lets go over both briefly.

Mans Laws

Contact your local Chamber of Congress or City Hall and ask them for operating permits and other information you would need to operate a business of this type from your home. Remember that you are not only operating a business for hire to control pests but to also provide the following services: Nutritional Seaweed Spraying, Natural Pest Control, Consultation, Organic Gardening Services, Drip Systems, Natural Tree Care , Natural Composting Services, Recycling Services, Natural Fertilizations and much more!

Natures Laws

Also remember that since you are not using any chemicals like DDT etc., a Chemical Operators License will not be required nor wanted by you. Insist that a special license be provided your company. Tell them what you do and how you do it and offer to help them.

Before you get to that point, you can operate safely without any special licenses as long as you operate your service as a gardener service and do not use any chemicals. Use only the natural materials. You should advertise as a gardening service that provides 100% Organic Services. Check with your local city for more information.

Here is a list to help you set up your own Natural Gardening Service

1..Get a Good Name for your business. Make sure that its says that its 100% Organic, etc. Do a DBA. See your local chamber of congress.

2..Develope a LOGO for your business.
3..Make Business Cards for your business.
4..Make Letter Heads
5..Get an ansering service to answer all your calls.
6..Get a P.O.Box as a mailing Address(don't say POBox say Suite on mailing)
7..Develop your own flyers about your business.
8..Before you start to advertise make sure you have the following equipment set up:

1   35 gallon Power Sprayer
Preferably electric or solar but a properly operating motor will do). Should have 100-150 ft of hose.

1   Backpack sprayer 2 gallons.
1 unit which contains various different tools you will need from plastic gloves to measuring cups and will carry your mixtures.

1   5 gallon unit which carries the days spraying formula concentrate(or you can have ready to use per customer).

You will find a small pick up truck will work fine, however a trailer can be used along with a car if a pick up is not available.

There are a few more items you will have to take care of for your business

You will need an area to work from, a quarter to one acre will do. This is because

you will be producing your own compost as well as setting up your own recycling center. An Ideal set up is to also have a store front from which to allow the public to buy your compost and other natural items. Check first to see if there is one already in your neighborhood. A shed will be useful also.

Another thing is that you should not attempt this by yourself. Ideally it should be done with several people involved. This is a great way for a co-op to generate their money or for a family, as women can do this easily also. This is also a great part time job for a student. You can start such a service near your college.

P.S. If you are a IGA member and you have decided to operate your own service and want special training then contact The Invisible Gardener for more details. Thank You....

Appendix 3

YOUR PET AND YOUR PLANTS, HOW SAFE?

"While it is true that animals can cause a great deal of damage to our Gardens, Plants can also be dangerous to our pets. Many plants around the garden are poisonous and can kill your dog or cat"

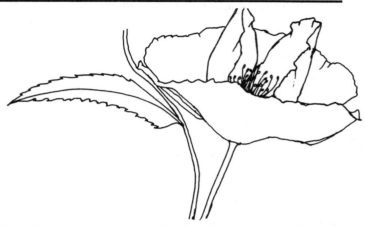

When we study Organics, we also study the Science of Plant Toxicity. This toxicity works two ways, it protects the plants from the insects attacking them, and also protects them from being eaten by animals. As gardeners, we must be aware that many plants are dangerous when injested. Small creatures such as cats, dogs and children are especially vulnerable. The young are in greater danger since they explore their environment using their mouths and hands.

It is not true that our pets will instinctively know what plants are ok to eat. If they were living out in the wild and had the benefit of a parent that showed them the way of the world from hunting to what plants not to eat, then they would probably not eat it, but with our modern day pets, many of Natures Secrets lie hidden from them. Exposing our pets to these various exotic plants can be a learning experience that they will never forget and hopefully they will develop the correct habits for living together with us.

page 123

Here is a list of some common plants that are toxic or have a bad effect on your pets:

Oleander (Nerium oleander) Every part of this plant is toxic. This plant will kill a grown man if he eats only one leaf!

Datura Highly fragrant and very poisonous. If the variety of Datura that you have has velvety leaves, this is a very poisonous variety that can cause damage just by rubbing against the leaves.

Castor Beans (Ricinus communis) One bite and your dead. Very toxic. The leaves are toxic also.

Azaleas leaves if injested in large amounts are deadly.

Rhododendrons leaves are toxic.

Boxwood(Buxus) all parts of plant toxic.

English Ivy (Hedera helix) fruit and leaves are toxic.

Heavenly Bamboo contains cyanide.

Larkspur(Delphinium) all species of Larkspur and Delphiniums are highly toxic.

Ornamental **Tobacco**(Nicotiana) contains nicotine.

Oxalis all species contains oxalates. Pigweed(Amaranthus spp.) contains oxalates.

Spurges(Euphorbiacea) gopher purge, poinsettia are euphorbias and highly poisonous. **Poinsettia** Can cause vomiting if injested.

Gopher Purge contains toxic amounts of oxalates that form crystals in the kidneys.

Some vegetables that are toxic to animals.

Spinach, rhubarb, potato vines, onions and tomatoes may cause allergic reactions to your pets or may even be fatal to them.

Bulbs such as tulips, daffodils, amaryllis and iris are dangerous if injested.

Apricots and **Peach** pits contain cyanide and your pets should not be allowed to consume them.

LEARN MORE ABOUT POISONOUS PLANTS:

"Plant Poisoning in small companion animals" by Murray E. Fowler, published by Ralston Purina Co.

Appendix 4

In our efforts to Reduce, Reuse, Recycle, Rethink and Replant, we are going to have to look at where our wastes are coming from and where they are going. Their "life Span" or their actual use time can be either very long or very short. Some things we will reuse over and over again and somethings can be used only once. For this article we will look at a source of incoming wastes...Junk Mail:

What is Junk Mail? Basically any mailing that you receive that you did

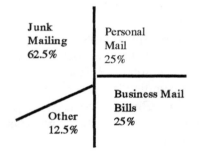

not request, asking you to buy, trade, sell, something you may not want or need: this does not count your normal mail from family, subscriptions, important letters such as billings, etc.

As you can see Junk Mail consists of 62.5% of our incoming mail. A great deal of this Junk Mail is looked at once then discarded. This has a very fast usage life span yet has a very long recycling life span.

What this means is that while you only use that mail for a short period of time (you read it once) it will take possibly from several months to several years to complete the cycle.

Here are several things you can do to reduce this amount of wastes both incoming and out going.

To Reduce incoming Junk Mail

there are several things you must do:

1....If you have any subscriptions then write to them in care of their mailing lists dept. Request that they remove you from their mailing lists which they have for sale or rental.

Tell them that you do not wish to have your name made available for this purpose and want it to be removed.
2....There are many organizations which deal with Direct Mailings.

JUNK MAIL JUNK MAIL
JUNK MAIL JUNK MAIL
JUNK MAIL JUNK MAIL
JUNK MAIL JUNK MAIL
JUNK MAIL JUNK MAIL
JUNK MAIL JUNK MAIL
JUNK MAIL JUNK MAIL
JUNK MAIL JUNK MAIL
JUNK MAIL JUNK MAIL
JUNK MAIL JUNK JUNK
JUNK JUNK JUNK JUNK
JUNK JUNK JUNK JUNK
JUNK MAIL MAIL MAIL
MAIL MAIL JUNK MAIL

Here are three:
JUNK MAIL BUSTERS
SUITE 5038,
4 Embarcadero Center
San Francisco, CA 94111

The Junk Mail Prevention Kit
K.D. Enviro-Ventures Inc.,
P.O. Box 30313,
Indianapolis, IN 46230

Direct Marketing Association's Mail Preference Service
11 West 42nd St, P.O.Box 3861,
New York, N.Y. 10163-3861

Write to them. You will find that this will remove close to 50% of your present junk mail.
3...You can reuse many junk mailings if you have friends that are interested in them, many companies want to sell their products and don't care who buys it. Also you will find that much of the junk mail that you receive will be made from recycled paper and can be recycled again and again.

Recycled paper can be composted provided that the following conditions are met (we will be discusing this and kitchen composting in our next issue in more detail):
1..Only use recycled paper that you know the ink is made from soy beans. This is the latest technology for paper. Also its OK to compost paper that is made from Hemp.
2..Do not use paper that has color or shiny surfaces as the inks have high levels of lead in them. To be sure, write to the newspaper company and ask them. Tell them that you are thinking of composting your newspapers and could they send you information on the inks, paper and any other compostion that you should know. They will love this! We asked our local newspaper to send us some information which we will print in our next newsletter for your information.
3...Use a paper shreder to tear up the paper before composting. We will go over the actual process of this composting in our next newsletter.

Appendix 5
The Garden Filter

"At last...an easy way to grow plants the organic way, help protect the environment and conserve water at the same time."

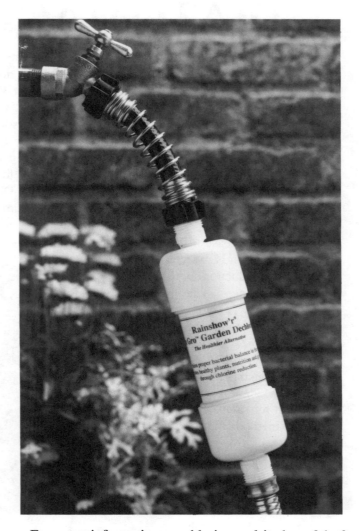

The health hazards associated with chlorine are becoming more known to the general public. Most cities in the USA and the world use chlorine in their water. Why? Chlorine kills bacteria, living organisms such as fungi, which contaminate our water source and our oceans. Chlorine does a great job of killing these organisms that can be harmful; but plants and the soil rely on some of these fungi and bacteria to function.

Live soil is the corner stone of the organic method. The soil must be alive with beneficial bacteria and fungi in order for the "system" to work. Food is eaten first by these bacteria then made available to the plants thru the soil. Every plants survival is based upon receiving nutrition from the inter-action between living organisms in the water and soil. Remove these 'friends' and we have big trouble! Many diseases originate from dead soil.

Many benefits of organic gardening are lost when chlorinated water is introduced into the eco-system of the garden. Natural spraying depends upon using clean filtered water to obtain maximum results. An organic gardener or professional does not always have the time to solarize the water so a good long lasting garden filter is important.

The Rainshow 'r Gardn' Gro filter uses a unique "redox" media of copper and zinc and works on the principle of electrochemical oxidation reduction. It converts harmful chlorine into a harmless chloride.

For more information on chlorine and its harmful effects read " Nature & Health "magazine, Vol. 10, no 4, summer, 1989. Written by John F. Ashton and Dr. Ronald S. Laura. Pages 44-49. Article is titled "One Hundred years of water chlorination".available from Whedon Young Productions, PO Box 170, Willoughby 2068, NSW Australia.

Until today most gardeners had no choice of the water they used in their gardens. But now it is possible for all gardeners to steer away from chlorinated water and move towards a cleaner, healthier choice with the Rainshow'r Gardn' Gro filter.

The manufactor can be reached at:
Pacific Environmental Marketing and Development Company, 421 S California St., Unit "D", San Gabriel, CA 91776 1-800-243-8775
If you are an IGA member you can call
1-800-354-9296   page 125

Appendix 6
DIATOMACEOUS EARTH

Over 400,000 tons of pesticides are applied each year by American farmers with less than one-tenth of one percent actually reaching targeted pests! A main source of contamination of our soil, water, air and food as well as being highly inefficient, this method of pest control is places at risk the health of the consumer and farmer alike.

With the increase of of organic methods of controls , we are seeing a reduction in chemical usage. The farmer of today is beginning to understand the delicate balance of nature of which he is directly responsible. With this knowledge comes the search for safer more effective methods of pest control.

Into the ring comes a long time favorite of organic farmers; DE. Thru out this book, I have been referring you to Garden Grade DE(not the pool kind). See DE page 18.

DE(Diatomaceous Earth) contains 14 trace elements, is non-toxic and biodegradable. DE controls via a mechanical action rather than chemical action, importantly, what this means is that subsequent generations of insects can not become resistant to it's effects.

DE Structure and it's Function

De is derived from dead sea life called Diatoms. There are over 1,500 species of diatoms and each has it's own skeleton shape. Most are snow flake shaped with a few being tubed shaped and some cone shaped. The shape is very important in it's effectiveness. Their composition will vary according to the area they lived and died in. Most DE contain silicon, sodium, aluminum, manganese, boron, magnesium, iron, calcium, and copper. The amount of amor-

phous(uncrystallized) silica the diatoms have percent during fossilization determines the amount of crystalline silica found in todays DE. The greater the concentration of crystalline silica the greater it's effect on insects. Also Fresh water DE deposits contain lower percentages of silica than does salt water deposits.

How does DE work?

Look at De under an electron microscope and you will see that DE crystals(crushed Diatom shells) have

very shape edges with a large surface area(making DE very porous). Amorphous DE is very different from crystalline silica where health is concerned. Crystalline silica is hazardous to ones health while garden grade DE(Amorphous). Crystalline silica causes Silicosis(the World Health Organization has declared that crystalline silica levels in agricultural grade DE exceeding 3% can be dangerous to humans). This is why pool grade DE is not safe to use on your garden or farm. Pool grade DE is made by super heating Amorphous DE until it becomes near 100% crystalline silica! Many companies add an acid to complicate matters. Avoid using Pool grade DE at all costs!

DE causes insects to loss their bodies water content by over 10%. DE effects the insects natural coating of cuticle(a waxy coating secreted by the epidermis). The sharpe edges of the DE particles lacerate this coating and penetrates in between the insects body plates, and absorbs any water(moisture) t comes into contact with. This slowly dehydrates the insect. The DE effects the insect from the outside as well as from the inside if the insect eats the DE as well.

DE can be fed to your dogs, cats, horses without any harmful effects to them. It is a natural dewormer while providing trace minerals.

A good formula to follow for feeding is:
One tablespoon per daily feeding for your larger dogs(over 50 pounds)
One teaspoon per daily feeding for your smaller dogs
1/2 teaspoon per daily feeding for your cat
One cup per daily ration for your horses.

DE can also be added to their drinking water at 1 tablespoon per gallon water or 1 cup per 50 gallons water.

By feeding DE to your animals, you are also providing for natural fly control since DE will reduce fly populations. This happens because DE travels thru the animals digestive system, and is deposited, unaffected, in the animals feces. The flies deposit their eggs as usual in the manure but few larvae will survive because they must move thru the manure in order to feed. DE can also be eaten by people. I suggest that you take a teaspoon in your meals once per week.

Grade grade DE is classified as GRAS(Generally Recognized As Safe) by the federal government and has been exempted from the requirement of residue tolerance on stored grain. It is so far be considered by most health professionals that Amphorus DE is not cancer causing in animals and humans.

A word of Caution

While DE is safe to use, it would be wise to follow a few safety rules:
1..avoid breathing, use a face mask if you have a breathing problem.
2..avoid contact with the eyes. Wash with water asap. Do not rub !
3..avoid over use. DE will kill beneficials as well as kill of earthworms if too much DE is applied to the ground.

While De does a great job of protecting plants from insect, it will do a better job if the overall coverage is increased. This is best done by using an Electrostatic Nozzle developed by PermaGuard of Albuquerque NM. This device charges particles by forcing them thru an ionized filed, where they pick up negative electrons. Surfaces that are attached to the ground are positive and will attract the DE dust. And due to the fact that like charges repel, when one area of a surface is covered, it repels on-coming particles, forcing them to move over and find an unused place. Eventually, the dust will cover the entire leaf surface, both on top and underneath. This makes for a very effective use of DE.

See index for more uses for garden grade DE

Some Sources of DE
DE distributors and manufacturers

Nitron Industries, Inc.
4605 Johnson Rd
P.O.Box 1447
Fayetteville, AR 72702-0400
1-800-835-0123

Peaceful Valley Farm Supply
P.O.Box 2209
Grass Valley, CA 95959
1-916-272-GROW

ARBICO
P.O.Box 4247
Tucson, AZ 85738
1-800-827-2847

Electrostatic DE Applicator

Perma-Guard, Inc.
3434 Vassar NE
Albuquerque, NM 87107

Appendix 7
SAFE SOAPS

NAME & ADDRESS	MADE BY	CONTACT NUMBER
Dr. Bronners Soaps P.O.Box 28, Escondido, CA 92025	All-One-God-Faith, Inc	1-619-743-2211
Shaklee's Basic H 444 Market St, San Francisco, CA 94111	Shaklee Corp available only thru local distributor	1-415-954-2007
Amway's LOC 7575 Fulton St. E, Ada, MI 49355-0001	Amyway Corp available only thru local distributor	1-616-676-7948
Citrus Organic Cleaner	Natural Bodycare	Available in Most Stores
Citra Solve	CitraSolve	Available in most stores
CitraChief Sacramento, CA 95818	Apache Enterprises	Available in most stores
Herbal Soaps PO Box 106 Altadena, CA 91001	Erlander's Natural Products	213-797-7004
Coconut Oil soap 3920 24th st San Francisco, CA 94114	Common Scents	414-826-1019

This is a list of only a few sources of natural soap products. Read the ingredients before buying. Experiment with the various organic soaps to use for pest control. Remember that you are defeating the purpose of growing organically if you use a soap with polluting chemicals in it.

Notice
Address and phone may and will probably change. We are not responsible for any such changes. If you are an IGA member, contact IGA for new address/ phone.

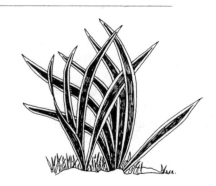

Appendix 8

PREPARATIONS OF FORMULAS

Garlic Juice from cloves.

Take cloves and crush(either run through juicer or use garlic press). Use 1 bulb(which has many cloves attached) per gallon of distilled water(or preferably solarized). Crush each clove and allow to sit in gallon water over night. Run through strainer. You can also make an infusion from the leaves.

Solar Tea.

There is an art to making solar tea. If you have ever made solar tea for yourself to drink then you basically know how to make this tea. It is called a tea because this is how it is made(with a tea bag). It can also be called an infusion. I prefer to call it Tea. Solar Tea is made by placing the ground up dried leaves of the herb into panty hose tied into a ball(becomes a tea bag) then place into a gallon(or larger) glass container of pure water(not city water, distilled or filtered water is best but stream water is ok). You should always use a water filter to control unwanted bacteria and toxins(chlorine). Allow to soak in the direct sun and moon light for 24 hrs. Pour into a clean dark glass container. Use within the next 24 hrs. The strength of this mixture will depend upon the purpose and the herb used.

What part of the plant are you using?

Leaves... Use dry leaves, grind in mortar and place in panty hose,tie into a ball and allow to soak for 24 hrs, (use large gallon (or 5 gallon)glass containers only, use only once.

Bark.. Allow bark to dry on sun tray, then grind bark to powder,place in panty hose, tie into ball and add to container of water and allow to soak for 24 hrs, strain through filter.

Seeds... Grind seeds with coffee grinder,place into panty hose, tie into ball and allow to soak for 24 hrs. Filter before use.

Flowers... Pick flowers, allow to dry on solar tray then grind into powder with mortar, place into panty hose and allow to soak for 24 hrs.

Fruit... the fruit can be either dried or used fresh. Dried.. place fruit on solar dryer, allow to dry slowly and grind to a fine powder. Place into panty hose to soak. When using fresh fruit, it is best to use a blender and liquify the fruit for easier application.

Root...can be used two ways, fresh or dried. when used fresh, can be boiled and a thick mixture can be made by allowing to soak for 24 hrs over a slow heat. When used dried can be added to panty hose to soak for 24 hrs.

SOLARIZATION OF WATER

When ever you are preparing a spraying formula, always use solarized or filtered water. Solarized water has been allowed to sit in the sun for several days. The mixture should be stirred clock wise for 5 minutes then stirred for five minutes the other way. This should be done once every day that the liquid is being solarized. Use a glass container if possible. Colored glass will change the energy level of the liquid. Experiement for best results. Green glass will be very helpful for promoting new growth on plants while yellow for new buds and blue for fruit. Red is useful for repelling insects.

HOW TO PREPARE HERBAL TEAS FOR SPRAYING

Method 1

Empty herbs into a large pot of boiling water. cover and steep for 1/2 hr. Pyrex or Corningware are preferred containers. No aluminum or cast iron. One tablespoon per cup of boiling water.Strain before using. Can be used straight as a spray etc.Excellent formula to use for repelling.

Method 2

Slowly bring water to a boil(use low flame), then turn off heat, add 1 cup of the herb, steep for 5 minutes, turn on heat to low and simmer for 5 min. Turn off heat and allow to sit over night.Use strainer to filter and pour into clean glass container. Label. Use within 24 hrs. Makes a very strong concentration. Excellent for making a fungus control spray or ideal for that hard to kill pest..

Making a Slurry

When you are using a dust or herb that is very dry, it will not blend well with water at first. Therefore a slurry is made. This is done by 1/2 filling a cup with the material you want a slurry of. Add 1 drop of Shaklee's Basic H or any other natural soap. Slowly adding water while stirring. The mixture will accept water and dissolve into a paste like substance. This can be then added to water and used as a spray. I suggest adding to 1 quart water, and dissolving by shaking well. Save this quart and use 1 cup per gallon liquid to be sprayed.

Appendix 9

The Natural Pool

Many of us seeking alternatives to chemicals who have pools have had to use chlorine and other pool chemicals, up until now that is. Now you can get a solar powered pool purifier.

Here's how it works

Each unit has small solar cells that produce low voltage (120 volt) electricity. The ionization unit uses the electricity to produce very low concentrations of silver and copper ions in the pool water. This amount is harmless to people and animals but lethal to bacteria and algae.

Each unit has a small copper/silver rods which sticks out the bottom into the pool. The suns light shines onto the solar panels and causes the rod to give off the ions into the pool water. Most units come with a kit which you use to measure the concentration level of the ions in your pool.

When it reaches the correct level(enough to kill the bacteria/algae in the pool water), you simply remove the unit from the pool. You will soon develop a cycle of times in and out of the pool. This will depend on where you live, usage of pool and other weather factors.

Using this sytem will help you to reduce the chlorine levels in your pool and given time and proper conditions to completely remove.

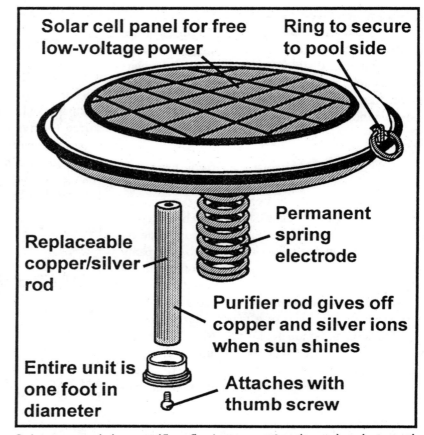

Solar cell panel for free low-voltage power

Ring to secure to pool side

Replaceable copper/silver rod

Permanent spring electrode

Purifier rod gives off copper and silver ions when sun shines

Entire unit is one foot in diameter

Attaches with thumb screw

Solar-powered ion purifier floats on water in swimming pool.

Steps to chlorine reduction

1...use solar ionizer as per instructions

2...use H2O3 instead of chlorine

3...use chlorine only when problem persits(algae/bacteria)

By following the above steps you will be insured that you chlorine levels will be reduced.

H2O3

Hydrogen Peroxide

H2O3 is a very good substitute for chlorine as a bacterial/algae control. You will have to experiment with the amount of drops to use. Use only food grade . I suggest that you try 10 drops per pool per week. Combine with the above solar unit for best results.

This until is currently availabe thru many mail order companies or you can try The Sharper Image catalogue

1-800-344-4444

ask for Floatron Pool Purifier

Terra Nova Ecological Landscaping owner Ken Foster (front) and employee Bill Kyte.

Tread Lightly and Pedal a Big Rake

You may see us sometime on the road, two landscape gardeners with all the tools of the trade—mower, trimmer, blower, rakes, broom, hoes, trash cans, etc. One thing will be noticeably missing: the truck. In its place the most efficient form of transportion—the bicycle! Terra Nova started the "Tread Lightly" service in September of '91 with one trailer for the tool box and lawn mower, the other trailer for the trash cans, blower, string trimmer, rake and broom. Between the two trailers we found we could carry everything we needed to perform professional landscape maintenance.

Our "Tread Lightly" service has been on the road now for over a year, and it has proven itself in that time. As far as we know Terra Nova is the first professional landscaping company in the state to use bicycle teams to expand their existing landscape business. You may have seen one of the articles in the local press about us. We have learned a lot in the last year and are confident we can provide quality service caring for our customers landscapes *and* care for the environment in a broader sense. The bicycles are one way of living up to our name, Terra Nova Ecological Landscaping. Beginning in January '93 we will have two new trailers on the road, serving the University Terrace area of Santa Cruz. The trailers are made by local trailer maker John Welch and signs painted by local artist Peter Bartczak.

If you have any comments or questions about our "Tread Lightly" service, please write to "New Earth News" P.O. Box. 677, Santa Cruz 95061-0677.

Reprinted from *California Bicyclist*

TERRA NOVA PROFILE

Ken Foster takes his environ-metally-sound landscaping business very seriously. He uses organic pesticides and fertilizers, elbow grease to get rid of pesky weeds, non-polluting electric powered tools —and "landscape trucks" fueled with pedal power.

"Using bicycles was a logical progression for the business," said Foster, 33, owner of Terra Nova Ecological Landscaping. "And we love to ride bikes anyhow." People often see Foster and his two employees pedaling their custom utility trailers around Santa Cruz. One trailer carries shovels, rakes and other gardening equipment and another carries a large trash can for hauling away clippings — which usually go into a local compost pile.

While most jobs are within a five mile radius of Foster's office, he says he has no trouble getting a good workout. A fully loaded trailer weighs about 200 pounds. "And in Santa Cruz, we have some major hills." Although he still uses his truck for longer-distance jobs, Foster says he'd eventually like to be 100 percent bicycle operated. He recently bought a new bike trailer that will run on a rechargeable battery, which will give less fit riders an extra push up hills.

"It's a challenge to make it work," says Foster of his bicycle fleet. "But I'm proud that we do make it work. To me, it's just the wave of the future." — B.H.

408-425-3514

The Refractometer

The Invisible Gardener is a pioneer in organics and here is a "tip from the frontier"! The hand Refractometer is well worth taking a closer look. It is a "handy" portable tool for a variety of agricultural uses from the backyard gardener to the professional. Indeed, the food industry is the chief user of the Refractometer. (Shouldn't it be the pest control professionals?–Andy)

Gardeners will find it an important tool in determining the amount of minerals present in your plants or produce. By determining these results, you can see what condition the plant is in and take advantage of your findings to increase the amount of minerals in the plant's soil composition.

A Refractometer can also help the grower to determine optimum pest control conditions in the soil and plant. The grower can use this tool in implementing an organic fertilization prog based upon the Brix readings. An organic foliar feeding program will show immediate results on the Refractometer. It has been found that the higher the brix reading the lower the pest activity. A Refractometer used in conjunction with a Radionics scanner will give a more complete picture to the organic professional in determining if the current program is optimal.

Brix is the unit of measurement taken by refractometer. The chart illustrated gives quite a complete breakdown of brix readings for a variety of plants. These readings will vary based on the time of year, type of plant, and location.

Experimenting with the Refractometer will yield many pleasant results. Try it on ornamentals, trees, flowers, anything that grows! Take it shopping with you and use it to pick your fruits and vegetables–pick a fruit to test (tell the grocer what you are doing and that you will buy your test sample even if the quality does not meet your standards) and maybe you will set a great example for the grocer to use when he goes shopping. The organic farmer/professional can get immediate field results.

Add sample

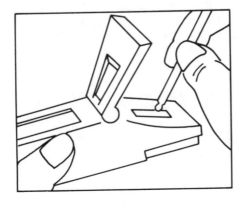

Close prism cover

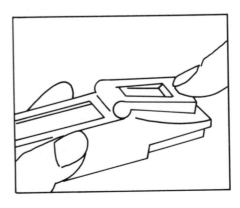

To use the refractometer, hold it in the horizontal position and swing the cover plate up to expose the measuring prism and the bottom surface of the cover plate. Be sure the exposed surface is clean.

Place a drop or two of the sample on the prism using the dipstick provided with the refractometer. Do not use glass or metal applicators to prevent scratching the prism. A plastic stirring rod is a good substitute.

Close cover plate ASAP. Hold the instrument for reading by placing the finger(s) on the cover plate and press the plastic cover gently, but firmly, as this spreads the sample in a clean, even layer over the prism. Point the instrument toward a light source and take a reading at the point where the dividing line between light and dark meets across the scale.

There you have it! You'll see that it is easy enough to use and if you need more information call me or contact me through TOBBS.

Special thanks to the folkks at Leica Inc for their charts and information. You can reach them at P.O. Box 123, Buffalo, NY 14240-0123

HAPPY GROWING,
ORGANICALLY, OF COURSE! IG

**Hold up to light
and read scale**

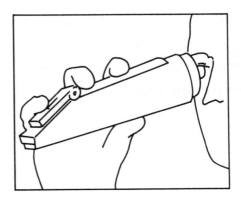

REFRACTIVE INDEX OF CROP JUICES CALIBRATED IN % SUCROSE OR ° BRIX

Refractometers are easy to use even for inexperienced personnel. To make a determination, place two or three drops of the liquid sample on the prism surface, close cover and point toward any convenient light source. Focus the eyepiece by turning to the right or left. Read percent sucrose or solids content on the graduated scale.

FRUITS	POOR	AVERAGE	GOOD	EXCELLENT
Apples	6	10	14	18
Avocados	4	6	8	10
Bananas	8	10	12	14
Cantalope	8	12	14	16
Casaba	8	10	12	14
Cherries	6	8	14	16
Coconut	8	10	12	14
Grapes	8	12	16	20
Grapefruit	6	10	14	18
Honeydew	8	10	12	14
Kumquat	4	6	8	10
Lemons	4	6	8	12
Limes	4	6	10	12
Mangos	4	6	10	14
Oranges	6	10	16	20
Papayas	6	10	18	22
Peaches	6	10	14	18
Pears	6	10	12	14
Pineapple	12	14	20	22
Raisins	60	70	75	80
Raspberries	6	8	12	14
Strawberries	6	10	14	16
Tomatoes	4	6	8	12
Watermelon	8	12	14	16
GRASSES				
Alfalfa	4	8	16	22
Grains	6	10	14	18
Sorghum	6	10	22	30
VEGETABLES				
Aspharagus	2	4	6	8
Beets	6	8	10	12
Bell Peppers	4	6	8	12
Broccoli	6	8	10	12
Cabbage	6	8	10	12
Carrots	4	6	12	18
Cauliflower	4	6	8	10
Celery	4	6	10	12
Corn Stalks	4	8	14	20
Corn, Young	6	10	18	24
Cow Peas	4	6	10	12
Endive	4	6	8	10
English Peas	8	10	12	14
Escarole	4	6	8	10
Field Peas	4	6	10	12
Green Beans	4	6	8	10
Hot Peppers	4	6	8	10
Kohlrabi	6	8	10	12
Lettuce	4	6	8	10
Onions	4	6	8	10
Parsley	4	6	8	10
Peanuts	4	6	8	10
Potatoes, Irish	3	5	7	8
Potatoes, Red	3	5	7	8
Potatoes, Sweet	6	8	10	14
Romaine	4	6	8	10
Rutabagas	4	6	10	12
Squash	6	8	12	14
Sweet Corn	6	10	18	24
Turnips	4	6	8	10

FOR REFERENCE PURE WATER HAS A READING OF "0"

Within a given species of plant, the crop with the higher refractive index will have a higher sugar content, higher mineral content, higher protein content and a greater specific gravity or density. This adds up to a sweeter tasting, more minerally nutritious food with a lower nitrate and water content and better storage characteristics. It will produce more alcohol from fermented sugars and be more resistant to insects, thus resulting in a decreased insecticide usage. Crops with a higher sugar content will have a lower freezing point and therefore be less prone to frost damage. Soil fertility needs may also be ascertained from this reading.

Resources Directory
Updated 7/10/94

This Resources Directory is updated several times per year. If you find a listing that is in correct heres what you can do....1 if you are an IGA member, call our club hotline and tell us and we will let you know the correct number and address. 2....if you are not an IGA member then I suggest that you send us a SASE along with the information on the incorrect addresses. We will return your SASE ASAP!(I wanted to do that!).

If you are listed and your address, etc., has changed then please send us your current info as we list all who send us their current info. If you wish to join IGA as a commercial member(Commercial members get free logo listing and 50 words in resources.) See IGA info/ordering at end of this book.

Happy Growing Organically of course!

The Invisible Gardener

ABUNDANT LIFE SEED FOUNDATION,

World Seed Fund
P.O.Box 772
Port Townsend, Wa 98368
206-385-5660

A non-profit corperation in the State of Washington whose purpose are to accuire, propagate and preserve the plants and seeds of the native and naturalized flora of the north Pacific Rim with particular emphasis on those species not commerically available, including rare and endangered species; to provide information on plant and seed propagation; and to provide information on plant and seed cultivation. They sell great heirlom seeds!

ACRES, USA,
P.O. BOX 9547,
KANSAS CITY, MISSOURI
64133   816-737-0064

A long time publication, headed by Chuck Walters. An excellent source of organic information and resources. Subscribe today and don't miss out! $15 per year. *Subscribe today!*

AFM ENTERPRISES, INC.,

1140 Stacy Court
Riverside, Calif. 92507
(714) 781-6860
low toxicity household products.

AGRI-GRO,

Living Waters Enterprises
HC4 Box 333
Doniphan MO 63935
1-314-996-7384
Contact: Ron Smith

Invented by Dr. Joe C.Spruill. Phd. Biochemistry. A biological complex deprived from natural compounds, processed through extration completed by fermentation. A plant and yeild simulant, contains stablizing nitrogen fixing bacteria, trace minerals, and humic acid.

Designed to improve chemical and fertilzer effecciency, improves natural sugar levels of plants, improves tilth and water capacity, desigend to reduce cost of production, reduce insect and disease problems and reduce soil compaction and also to reduce nematodes and salt buildup in soil.
An IG of A member. Tell Ron the Invisible Gardener sent ya!!

Sustainable Environmental Alternatives™

ARBICO,
ARIZONA BIOLOGICAL
CONTROL, INC,
P.O.BOX 4247
TUCSON, ARIZONA,85738
1-800-827-BUGS
1-602-825-9785

Sustainable Environmental Alternatives for the traditional, transitional and Organic grower.
An IG of A member

BARGYLA RATEAVER,

9049 COVINA ST,
SAN DIEGO, CA 92126
A leader in the field of Organics, a good source of ground kelp.

BAU,

P.O.BOX 190,
ALTON,NH 03809
603-364-2400
NATURAL WALL PAINTS, THINNERS, CLEANERS ETC

BIO-DYNAMIC FARMING & GARDENING ASSOCIATION., INC.

P.O.BOX 550,
KIMBERTON, PA 19442
215-935-7797
Your source for bio-dynamic information.

BIO-GARD AGRONIMICS, INC.,

P.O.BOX 4477,
FALLS CHURCH, VA
22044-0477
1-800-673-8502
CALCIUM-25
PRODUCTS

BIO INTEGRAL RESOURCE CENTER BIRC ,

PO Box 7414,
Berkeley, Ca 94707
415 524-2567
They are the leaders in information on Integral Pest Management.

BIO PAK ,

Diversified Packaging Products
1265 Pine Hill Drive
Annapolis, Maryland 21401
301-974-4411
Catalogue offers recycled paper products for businesses.

BOUNTIFUL GARDENS,

ECOLOGY ACTION,
5798 RIDGEWOOD RD,
WILLITS, CA 95490
707-459-6410

BOZEMAN BIO-TEC,

1612 GOLD AVE,
BOX 3146,
BOZEMAN, MT

CENTER FOR SAFETY IN THE ARTS,

5 Beckman St.
New York, NY. 10038
(212)227-6220
Safe Art materials.

CONSERVATREE PAPER CO.,

10 Lombard St, Suite 250
San Franscisco, Ca 94111
(415)-433-1000 1-(800)-522-9200
A national distributor of recycled paper products. For businesses . Offers office and printing papers.

CO-OP AMERICA,

1850 M ST., NW,
SUITE 700,
Washington, DC 200035,
1-800-424-2667
202-872-5307

CUSTOM COPPER,

P.O.Box 4939,
Ventura, California 93004
805-647-1652
They make various copper products which work against snails, slugs and other crawling pests.

EARTH SAVE,

706 FREDERICK ST,
SANTA CRUZ, CA 95062-4069
408-423-4069

EARTHWORKS PRESS,

Box 25, 1400 Shattuck Ave,
Berkeley, Ca 94709
(415)527-5811
Publishes" 50 Simple Things You Can Do to Save The Earth" by Earthworks Group.

DOWN TO EARTH

DOWN TO EARTH

800 Park Way
Lake Elsinore, CA 92530
909-674-8558
Ask for Danny
Distributer of Earth wealth Rock dust and Bio-Dynamic Compost.
An IG of A member

ECO-HOME,

4344 Russell Ave.,
Los Angeles, Ca. 90027
(213)-662-5207
visit her demonstration home on
recycling, low water plants,
Solar, Workshops, classes etc...
ask for Julia Russell.
An IG of A member.

The Carbon Atom

ECOLOGICAL BIOLOGICAL DYNAMIC ORGANICS, GROWING CRAZY ORGANICALLY,
1971 WESTERN AVE, #172,
ALBANY, NY 12203

Karl is an IG of A member.

ECO-NET NEWS,
3228 SACRAMENTO ST,
S.F., CA 94115
415-923-0900

ECOSAFE PRODUCTS, INC.,
P.O.Box 1177,
St Augustine, Fla 32085:
(800)-274-7387
For years they have been making safe 100% pure pyrethrum products for you and your animals. ZAP

ECO-VILLAGE
CRSP
Cooperative Resources & Services Project.
3551 White House Pl.
Los Angeles, CA 90004
213-738-1254
An IG of A member

EDUCATIONAL COMMUNICATIONS
P.O.BOX 35149
LOS ANGELES, CA
90035-9119
310-559-9160
NANCY PEARLMAN
Your guide to the worlds Environmental Crisis.
An IG of A member

ENVIROSOL,

ORGANIC PRODUCTS INC,
10430 MOORPARK ST,
SPRING VALLEY, CA 92078
619-670-1042
A good source of organic supplies.

FERTRELL CO.,

P.O.Box 265
Bainbridge, Penn 17502
717-367-1566
Makes Organic fertilizer blends.
Excellent!

FLEA BUSTERS,

213-470-FLEA
818-787-FLEA
714-979-FLEA
800-338-FLEA
619-297-FLEA
Natural Flea control services.

FULLY ALIVE RESOURCE GUIDE,

P.O.BOX 31232,
SANTA BARBARA, CA 93130

FILLMORE SNAIL FARM,

2041 W. Young Rd.
Fillmore, Ca., 93105
714-781-7643
Rumunia Decollata Snail Predator

FRIENDS OF THE TREES,

P.O.Box 1466
Chelan, Wa 98816

FOOTHILL AG. RES., INC.,

510 West Chase Dr.
Corona, Ca., 91720
714- 371-0120
Rumunia Decollata

GAIA,

P.O.BOX 1655,
EDMOND, OK 73083-1655
405-282-2086
Agricultural consulting services for ecological agriculture. George Kuepper

GARDENS ALIVE!,

Natural Gardening Research Center
5100 Scheneley Place,
Lawrenceburg, IN 47025
1-812-573-8650
They operate Gardens Alive gardening club and publish an excellent newsletter as well as provide you with many natural products.

GARDENER'S SUPPLY,

128 Intervale Rd.,
Burlington, VT 05401
802-863-1700
Provides information and products on Organic pest control methods, natural lawn care and moving towards the zero waste household.

GARDEN-VILLE,

7561 E. EVANS RD,
SAN ANTONIO, TX 78266
512-651-6115
MALCOLM BECK'S STORES
In business in 1952!

**GARRET, HOWARD,
THE NATURAL WAY,
P.O.BOX 140650,
DALLAS, TX., 75214
214-920-1838**

*Howard can be heard every Sunday
on his radio show in Dallas Texas,
WBAP(820 AM), 8 am to noon.*

**HARMONIOUS
TECHNOLOGIES
P.O.BOX 1865
OJAI, CA 93024
805-646-8030
805-646-7404 FAX
John Roulac
HOME COMPOSTING
SPECIALISTS**

Has assisted more than
100 cities, counties and
regional districts in
implementing home
composting programs.

Publishes "BackYard
Composting"......$6.95

**HOUSEHOLD HAZARDOUS
WASTE PROJECT,**
901 S National
Box 108
Springfield, Missouri 64804
(417)836-5777
They publish Guide to Hazardous
Products around the Home.

HUMA GRO PRODUCTS,
1 N Roosevelt
Chandler, AZ 85226
602 961-1220
Multi-purpose Fertil-humus,
humua-N, Huma-Iron and Thatch. Safe
non-toxic fertilizers . Contact Robert
Walker for more information.

**HEAL
HUMAN ECOLOGY ACTION
LEAGUE,**
P.O.Box 49126
Atlanta, Georgia 30359
(404)248-1898/435-8632
Heal provides many different support
services for chemicaly sentsitive people.

THE INVISIBLE GARDENER,

29169 Heathercliff Rd 216-4O8
Malibu, Ca 9O265-4146
310-457-4438, 1-800-354-9296

TOBBS Computer System:
310-457-4268 24 hrs 2400 8-N-1
The INVISIBLE GARDENER'S OF
AMERICA(IG of A),
ORGANIC SUPPLY HOUSE,
ORGANIC PEST CONTROL
SERVICES, ORGANIC SPRAYING
SERVICES, ORGANIC
CONSULTATION, ORGANIC TREE
CARE. Newsletter on growing
organically, publishes this book
(free to members), Makers of
Organa (tm)- Super Compost,
Malibu Gold(tm)-100% Pure LLama
Manure, Organa Gold,
SuperSeaweed(tm) and more
organic stuff like that there!
Send $3 for latest newsletter! $35
per year for membership, $25
seniors(with proof of age), $25
non-profit(ok trade membership) or
$100 for commercial. Andy can be
heard every third Monday KPFK
90.7 FM in LA. 3:00pm to 4:00 pm.

Earth Wealth
1491 Daffodil
Canyon Country, CA 91551
805-298-0791
ask for Walt Zmed
Rock Dust

MAXICROP USA,
P.O.BOX 964,
ARLINGTON HEIGHTS,
ILL, 60006

NATIVE SEEDS/SEARTH,
385O West New York Dr.
Tucson, Az 85745
Heirloom seeds

NATIONAL COALITION AGAINST THE MISUSE OF PESTICIDES,
530 7th St. SE
Washington, D.C. 20003
202-543-5450
Publishes "Pesticides and You".

NECESSARY TRADING COMPANY,

328 MAIN ST,
New Castle, Vir 24127
703-864-5103
A very large organic supply house.

NORTH COUNTRY ORGANICS,

P.O. Box 1O7
Newbury, Vermont O5O51
8O2 222-4277

NITRON INDUSTRIES, INC,
4605 JOHNSON RD.
P.O. BOX 400
FAYETTEVILLE,
ARK 72702
800 835-0123

A natural product company for organic growers. Many useful items for pets too. Dustin-mizer, septic tank formula, seeds. Makers of Nitron A-35!! Also sells Diatomaceous Earth, liquid seaweed, organic admendments.
CALL FOR FREE CATALOG. TELL THEM THE INVISIBLE GARDENER SENT YA!!
An IG of A member.

PEACEFUL VALLEY FARM SUPPLY,

P.O.Box 2209,
Grass Valley, Ca 95959
916 272-Grow

A complete organic growers catalogue. Natural pest controls, organic fertilizers, nursery stock, seeds and bulbs; hand power tools, consultation, natural pest management.
An IG of A member.

PRANA,

SEABRIGHT AVE,
SANTA CRUZ, CA
95063-2335
408-462-2762
LIQUID SEAWEED

Environmental Printers

PRECISION

PRINTING & DESIGN

PRECISION PRINTING
14544 KESWICK ST.
VAN NUYS, CA 91405
818-988-9990
FAX- 818-988-9996
ENVIRONMENTAL PRINTERS
CONTACT PERSON:FLOYD

Environmentaly Friendly printer using soy ink, recycled paper, safer printing chemicals.

An IG of A member

PROJECT RAINFOREST,
3900 FORD RD,
PHILADELPHIA,
PA 19131
215-473-5131
BRUCE SEGAL

TAKE A TRIP TO A RAINFOREST WITHOUT LEAVING YOUR CLASSROOM!

AN IG OF A MEMBER

REAL GOOD NEWS,
966 Mazzoni St.
Ukiah, Ca 95482
1-800-762-7325,
1-707-468-9214

The Largest and most through selection of alternative energy products in the world! Also carries a variety of natural products.

RINGER,
9959 VALLEY VIEW RD.,
EDEN PRAIRE,
MN 55344-3585
1-800-654-1047

RONNIGER'S SEED POTATOES,
Route 3, Moyle Springs,
Idaho 83845
Organically Grown seed potatoes
send $1 for catalog

R-VALUE,
P.O.Box 2235,
Smyrna, GA 30081
800 241-3897
makers of Drax(tm) 5 % boric acid

SAFER SOAP,
see Ringer

SANTA FE NATURAL TOBACCO COMPANY,
P.O.Box 25140
Santa Fe, NM 87504-5140
505-982-4257, Fax- 505-982-0156

Sells smoking tobacco grown organically without chemical additives. Can also be used for pest control, teas, etc. To be used with Tree vents.

SEEDS OF CHANGE,
621 OLD SANTA FE TRAIL #10,
SANTA FE , NM, 87501
HEIRLOM SEEDS.

SEED SAVERS EXCHANGE,
Rural Route 3, Box 239
Decorah.la 52101

SHAKLEE CORPORATION,
444 Market St,
San Francisco, CA, 94111
1-800-SHAKLEE
Basic H soap. Call and ask for local distributor.

SHEPERD'S GARDEN SEEDS,
30 IRENE ST.,
TORRINGTON, CT 0
6790
203-482-3638

SHURE CROP BRAND,
Made by Hi-Bar Ltd.
P.O.Box 2075
1825 Kusel Road
Oroville, Ca., 95965
916-534-7603
This product stimulates biological organisms to grow. Unlocks trace minerals, etc. excellent foliar feeder-100% Organic.

TREE PEOPLE,
1260 MULHOLLAND DR,
BEVERLY HILLS, CA 90210
818-753-4609

WILDERNESS INSTITUTE,
2818 AGOURA RD,
AGOURA HILLS, CA 91301
818-991-7327

WINDSTAR,
2317 SNOWMASS CREEK RD,
SNOWMASS, CO 81654
1-800-669-4777

ZEROCHEMICS,
1005 WEST HICKORY,
JERSEYVILLE, IL 62052
1-618-498-9322
RON SEARS
NIGHTCRAWLER FARMS,
WORMS AND CASTINGS.

NOT LISTED?
CALL US!
1-800-354-9296
1-310-457-4438 office
The END(for now).

HEAL THE EARTH

AND

YOU HEAL YOURSELF

RESOURCES

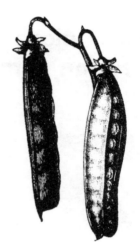

RESOURCES

When ordering, tell them The Invisible Gardener sent ya! Ask them if they will give you a 10% discount for being an IG of A club member.

Earth Wealth
1491 Daffodil
Canyon Country,
CA 91551
805-298-0791
ask for Walt Zmed

Harmonious Technologies
The Home Composting Specialists

P.O.BOX 1865
OJAI, CA 93024
805-646-8030
805-646-7404 FAX
John Roulac

Nitron Industries, Inc.

4605 Johnson Road • P.O. Box 1447
Fayetteville, AR 72702-1447
Phone TOLL FREE: 1-800-835-0123

Organic Headquarters
...to make your Garden GROW!

PEACEFUL VALLEY FARM SUPPLY
1-916-272-GROW
They have moved into their new warehouse in Grass Valley. Call for Catalog.

page 146

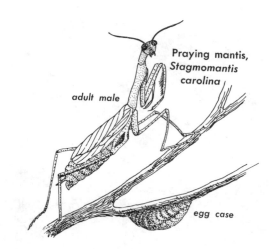

Praying mantis,
*Stagmomantis
carolina*

adult male

egg case

Giovanna De Santi Medina
Editor, PPNF Journal

INDEX

R

S

T

POLLUTION BEGINS AT HOME

Pollution does begin at home. Our shopping power causes changes throughout the world. To stop chemical pollution we need to stop buying chemicals that contribute to this problem. By following the 5 R's we can change the way people live and work and their relationship with the earth.

Reduce, Reuse, Recycle, Rethink, RePlant.

We can begin by reducing our chemical dependency. This is done thru careful Rethinking, Recycling and Reusing our available resources.

The Invisible Gardener's of America operates a 24 hrs Online BBS for the following purposes:
1..to make information available thru the fastest means available to us.
2..to help you reduce by not having to drive(save gas, etc); not having to send mail(save paper, stamps,time etc); and by not having to waste more of your time.

This service is available free to IGA members. See Information on Membership.
The Invisible Gardeners of America was started by The Invisible Gardener and promotes his knowledge in organics. Thru **TOBBS (The Organic BBS)**, The Invisible Gardener has made himself available to the millions of people around the world who need him and his expertise.

With this in mind he has developed **TOBBS**. Thru **TOBBS** people from all over the world can communicate directly with The Invisible Gardener, access Internet as well as local **E-mail**, access the **Organic InfoBank** as well as contact IGA club members from all over the world.

TOBBS provides quick access to information via your computer and a modem. With a modem hooked into your computer and also hooked into your phone line, you are ready to access the"Cyber World". You will also need the proper communications software to get online with. There are many pro-

grams available for you to use. TOBBS provides you with a special terminal program called Ripscript. This allows you to use the mouse in selecting your menus. TOBBS has been especially designed to operate on Ripscript. If you are an IGA member and you wish to get ONLINE with **TOBBS** you will first need a communication software to run on your computer and modem. Once ONLINE TOBBS you can obtain Ripscript free(after registration is processed(within 24 hrs)).

TOBBS is free to all current IGA members. Your only cost is your phone call. We will be upgrading that in the near future to allow you to make local calls to access **TOBBS**(ask us how you can do that).

Here is a listing of services and information available on **TOBBS**

The **Organic InfoBank** provides you with instant access to information on what ever gardening subject you want! From Organic rose care to organic lawn care to growing your own, Organic tree care and much more! All available thru our unique Online Rx located in the Organic InfoBank. Search and Retrieve provides you with access to information in the InfoBank and allows you to view Giff files associated with that information as well as Fax it to yourself! Thru the InfoBank you can access the internet via World Link and Chat Link! Exchange gardening ideas, questions, info, with millions of people around the world!

Online Classes

TOBBS provides you with Online Classes and workshops. Now you can take a course on Organic Rose care directly from The Invisible Gardener or you can choose an available instructor / course.

TOBBS provides online workshops / conferences for groups, organizations, professionals as well as for its members. Cost information on classes and workshops, etc., is available thru TOBBS.

Thru TOBBS, you will be able to get advice directly from **The Invisible Gardener**, Free! This book is available on **TOBBS** along with our **Online Newsletter**! We also have an **Online Newspaper** to keep you up to date(we are always looking for online reporters).

Take a visit to our **Online Compost Pile** or look at our **Online Gardens**! Go shopping with our **Online Shopping Mall**! Visit on **Online House**, and check out how to deal with Termites, Ants, and more!

All of this is available **FREE** to IGA(The Invisible Gardener's of America) members. **TOBBS** is on 24 hrs everyday for your service.

What is a BBS?

A BBS is an electronic Bulletin Board System Requiring a Computer and a modem hooked up to your phone line.

What is a Modem?

A modem allows signals to go between your computer and the BBS system and between your modem and the phone line.

Cost on TOBBS

TOBBS is FREE to all current IGA members.
There are however costs for certain sections/services of TOBBS. These are posted thru-out TOBBS.
Some examples are:
Online Classes usually cost $20 per course ·
Online Workshops usually cost $20 per workshop
Special Online Consultations are available by appointment

USING TOBBS

STEPS YOU TAKE TO GET ON TOBBS:

1..Make sure you have a terminal software to run on your computer. There are many available on the market. Choose one that has ripscript capablities.
2..Be also certain that you have a modem hooked up. Currently TOBBS uses only 2400 baud modems but this will be upgraded in the near future to 14,400
3...Dial into TOBBS from your terminal program. Use the setting as follows:8-N-1, Ansi, Full Duplex, XON/XOFF=OFF, RTS/CTS=ON,Auto-LF=OFF, BS=Destructive
4..Follow Online instructions
5..First Time Online type "NEW", Give membership number when asked for password(you can later change this).
6..Enjoy!(see you ONLINE!)
7...leave me Email and say Hi or question.
Happy Happy :)
Mr IG

310-457-4268

Andy Lopez(The Invisible Gardener) can be reached thru Internet at the following address if you have a question about TOBBS›

WL-Mr_IG@Society.com	Internet address
72775,1161	Compuserve address
Mr_IG @IGA	WorldLink address
Sysop or Mr_IG	TOBBS address

Alternatives to Chemicals for the Home, Garden, Farmer and Professional.

CLASSES WORKSHOPS AND RADIO SHOWS

AVAILABLE FROM THE INVISIBLE GARDENER

✣ LEARNING TREE

ENVIRONMENTAL GARDENING

More info:Learning Tree University
(805) 497-2292

✣ TREE PEOPLE

Andy gives a free workshop once a month on a Sunday, 10-1 at their Organic Test Gardens.
For reservations: (818) 753-4609

✣ SANTA MONICA COLLEGE

ENVIRONMENTAL GARDENING
A six week evening course

For info: Santa Monica College
310) 450-1145

✣ NITRON

Andy is a speaker at the Nitron conference every year in Fayetteville, Arkansas. This is one conference you don't want to miss!

For info: Nitron Industries
(800) 835-0123

✣ KPFK RADIO SHOW

HEALTH FROM THE GROUND UP
You can hear Andy on KPFK 90.7 FM in the L.A. Area every third Monday from 3:00-4:00pm.
Call in and Andy will answer your questions the best he can given the nature of a radio show with limited time available. The phone number to call is (818) 985-5735.

Do you want to hear Andy on every week? Give KPFK a call at (818) 985-2711 and tell them so.

✣ TOBBS

TOBBS ONLINE CLASSES
From your own home computer and modem. Each class is $20 and has four lessons. These are designed to take you beyond the consultant and interactive levels in the Bulletin Board System. Leave a message in E-mail to Andy if you're interested.

call 1-800-354-9296

and have The Invisible Gardener come to you area!

HEALING THE EARTH

"Man did not weave the web of life. He is merely a strand in it. Whatever he does to the web, he does to himself."

Chief Seattle 1856

By Andrew Lopez
The Invisible Gardener

The Healing process begins within ourselves. First we must understand our relationship to the earth for only in healing the earth can we truly heal ourselves. We are all linked toegther like a spiders web, what effects the part, effects the whole.

We cannot allow ourselves to fall into the trap that a false logic dictates. What is this false logic? To create and to function outside of nature. If you look closely at all things upon this planet, you will see that everything has its limitations and must therefore function within those limitations. Anything that functions outside of these limitations is alien to the world and a danger to us all.

We must understand that science must work within the laws of nature and not outside of it. To create outside of natures laws is to promote 'death'. For in the end, all things must die and revert back to its orginal form or source-Nature. Few animals upon this planet have been given the knowledge that mankind has. Animals are not aware of the universe, of atoms, of energy beyond limits. This is a special gift, let us not mis-use it.

As mankind, create and destroy we must, but we must temper this power, this knowledge, with the wisdom that we have also been given. We must allow Nature her rightful place along side us. We must see beyond our needs and provide for the needs of the whole if we are to survive.

We must set limits upon our growth, upon our powers. We can not achieve independence from Mother Nature, to try is genocide. Look to the whales, seek their help, seek help from all our planetary brothers and sisters. Lets Live closer to God, in Peace, in Harmony.

100% Organic

We provide the following services:
Natural Spraying Consultant
Natural Pest Control Specialist
Organic Tree Care Specialist
Organic Consultant to Professionals,
Corporations, Farmers and Gardeners.
Converting to Organic?
Speaking Services(we travel to you any where in the world!)
Workshops
Classes
Organic Training
Serving you since 1972

1-800-354-9296
"There are no problems only Solutions"

This book is written as the first step towards this goal. By reducing our dependence upon chemicals and increasing our dependence upon organics, we have a chance that our childrens children will see trees and know a fruitful earth.

With the help of this book, you can finally control ants without the chemicals, without killing yourself in the process. With the help of this book, you will be able to grow your vegetables totally organically. With the help of this book we will find our "Green Acres". Please use this book. Let us know what works for you. We are constantly expanding this book with each printing. If you cannot find something that you are looking for write to us.

Yours

Super Seaweed™

For all plants • 100% Organic • 5 drops per 1 gallon

A NATURAL LIQUID SEAWEED CONCENTRATE

SUPER SEAWEED (TM) is a very special Biodynamic prepatation which enhances the soil/plant relationship. This allows for more nutrition to be make available to the plants, thus less fertilization is required. contains over 70 trace minerals and many beneficial Bacteria and Enzymes and Humic acid all from 5 different seaweed sources throughout the world! All that we add to the seaweed is Rock Dust and spring eater. To be used as a foliar feeder. Use only 5 drops per gallon. Can be used with Nitron (TM), Agri-Gro(TM) and Shure Crop(TM) or any other natural fertilizers and products. Allows water to ecome wetter, allows more nutrients to be available to plants and soil, re-mineralizes the soil, planet. Helps trees to regain strength, increase root systems, helps Oak, Pine, and sensitive trees to recover when applied either as a deep root feeding or thru foliar. Will not damage trees. When using chemical fertilizers, use 1/3 less. Excellent as a compost starter, for fruit trees, roses, exotics, lawns. Use with inline feeding systems and hydroponics.

1 oz....$ 10.00 (makes 150 gallons) add $2.00 for shipping
8 oz....$ 40.00 (makes 1250 gallons) add $4.00 for shipping

ANDY LOPEZ

Living in the secluded hills of Malibu Canyon is a man known as the "Invisible Gardener." With a title like that, some confusion is inevitable: Could he be a playful, elf-like man who comes down from the canyon at dawn to sprinkle dew over the ground before others awaken? Or simply a wonderfully unobtrusive groundskeeper who literally fades into the landscape while dutifully taking care of the foliage? Actually, as Andy Lopez explains, *he* is not the "Invisible Gardener" at all — nature is. "I am just one of her helpers," he states. "And if I could get more people to do what I am doing, instead of consistently destroying the environment, she would have a much easier job."

Growing up in Puerto Rico, Lopez was heavily influenced by the fact that his mother grew her own fruits and vegetables and always used animal manure as fertilizer. This organically based philosophy toward gardening techniques laid the foundation of Lopez's beliefs, and in 1972 he founded Astra's Garden, based on something of a religion that subscribes to living in harmony with the environment: not polluting; treating all living plant life with respect; and, basically, just listening to what the earth is telling us.

People call him a "soil psychologist," yet he prefers the "plant nutrition specialist" appellation instead. Still, he does make an effort to get to know his clients — to get a sense of their lives — as much as possible. "I try to deal with the owner and his/her property as one entity," he maintains, going on to state that it is remarkable how much one's property reflects one's emotional state. His wide variety of clients — including such celebrities as Olivia Newton-John and Mark Harmon — attests to his effectiveness in connecting with people who share his motto ("happy, healthy, holy") when it comes to living and interacting with the natural surroundings.

Lopez is also the founder of the Gardeners of America club, which he started in an effort to raise the consciousness of all those concerned with their environnment. The club publishes a newsletter every 30 to 60 days which discusses new products, new procedures, and a number of timely topics relating to planting and growing. It also produces a constantly updated compendium of current environmental reports, providing an essential tool toward understanding the battle against — and alternatives to — polluting chemicals.

Lopez professes to be the only "absolutely organic" spraying specialist in the United States, pointing out that, while others might be "semi-organic," they all still rely on chemicals in one form or another. "If it doesn't come from mother earth, then it isn't organic," he states flatly. He goes on to say that 75 percent of all pollution in the U.S. comes from the home: "If people would go back to the basics and grow their own food, and take care of their bug problems naturally, chemical pollution would no longer be a problem."

Incorporating the philosophies of his various organizations and activities, Lopez's mission is the education of the public on the necessity to quit using chemicals altogether and, in the meantime, to properly dispose of their waste — so that one day the earth may return to its natural cycle of growing and decomposing. He also professes hopes of expanding his club internationally, noting that, "After all, these same problems exist all over the world; they just exist in varying degrees."

All Natural
PEST CONTROL

Andy Lopez is world recognized as an environmentally concerned horticulturist and has been in the field of organics for more than 20 years. Founder and CEO of THE INVISIBLE GARDENER ENTERPRISES, Andy has developed unique organic products and systems to naturally sustain your home and garden horticulture in perfect harmony with the environment.

Your Membership Benefits Include:

- NATURAL PEST CONTROL ALTERNATIVES, a book which shares the secret formulas to turn your herbs and spices into amazing remedies.
- Two newsletters a year which provide indispensible info to make gardens flourish.
- An 800 number to the ORGANIC HOTLINE (Includes 1/2 hr. free phone consultation).
- Unlimited phone consultation on the Club phone line.
- Immediate help on pest control problems & composting.
- Information on where to get products and services anywhere in the world.
- Member rates on classes, workshops, audio tapes, videos.
- Natural pest control, spraying and tree care services.

Call or write for compost and fertilizer products, tree vents, and biodynamic products. Send $3 to receive our current newsletter (or for immediate response call TOBBS).

The *Invisible* Gardener

310-457-4438

29169 Heathercliff Road
Suite 216-408
Malibu, CA 90265-4146

- -

Cut this page out from book and
return to us for processing.
NPC Code

Membership
Application
and Order Form

The Invisible Gardener
of Malibu

The Invisible Gardener says

"DON'T PANIC
IT'S ORGANIC !

The Invisible Gardeners of America
29169 Heathercliff Rd. #216-408,
Malibu, CA 90265-4146   310-457-4438 Office
310-317-2090   24 hrs Voice Mail
310-457-4268 BBS   Fax   310-457-2589

DATE:

Name _____

Address _____

City, Zip _____

Area code/Phone   (   ) _____

Phone(fax) _____
Please fill out and return with proper
funds: check or M.O. only please.

"Natural Pest Control Manual"
(this book is free when joining)

$19.95 plus $2.00 S & H per book
or $3.50 first class

Please check Box(s)

☐ I wish to order another book. Enclosed is my $19.95 plus $3.50 s&h plus $1.50(ca) tax

☐ **Yes!** I want to Join! Send me my 2 Newsletters per year and start helping me with my pest problems. If I already have your latest Book, I understand that I can join IGA with a $10 discount. My receipt for the book is enclosed.

☐ I want to become a commercial IGA member. Enclosed is my membership fee minus discount (If I allready have the book). Please find information about my product(s) or service(s). I understand if I am not approved I will receive a complete refund.

☐ No, I do not wish to join at this time. Send me your next newsletter. Enclosed is $3.

☐ Double Yes! I own a computer and I want to get online with TOBBS(The Orgnic BBS) FREE! I understand I will pay only for the phone tolls. Send me more info. Use my Fax# if available, otherwise heres a SASE.

```
            *** *** *** ***
Yearly Membership Fee(as of 8/94)
Individual......$35        _____
Non Profit.......$25        _____
Senior Citizen  $25        _____
Commercial....$100.00      _____
Book($19.95 +$3.50 s&h)$_____
tax on book (ca only)        $1.50
Total  Inclosed          $_____
```

Please note as of 1/95 the following new prices are in effect:
New members...........$55 1st year (book)
Renewals .... $25 per year(no book)
Seniors and Non-profit....$35 1st year(book)
Commercial membership......$100 per year

☐ Non Profit Organization

☐ I am a senior citizen , here's proof of my age

Any questions? Call 1-800-354-9296
Note :This order form replaces all previous forms as of 8/94

NOTES

TRAVELING
TO
TONDO

A Tale of the Nkundo of Zaire

Retold by VERNA AARDEMA

Illustrated by WILL HILLENBRAND

DRAGONFLY BOOKS • ALFRED A. KNOPF
NEW YORK

To my great-grandson Nicolas Adsit,

who arrived before this book

V. A.

For Joshua

W. H.

A DRAGONFLY BOOK PUBLISHED BY ALFRED A. KNOPF, INC.

Text copyright © 1991 by Verna Aardema
Illustrations copyright © 1991 by Will Hillenbrand

Library of Congress Catalog Card Number: 90-39419
ISBN: 0-679-85309-X
First Dragonfly Books Edition: January 1994

Manufactured in Singapore
10 9 8 7 6 5 4 3 2 1

GLOSSARY and GUIDE TO PRONUNCIATION

(in order of occurrence)

Tondo (TON-doh): A town on Lake Tumba

Nkundo (uhn-KOON-doh): People who live in the rain forest of Zaire; their language
    is called Lonkundo (Lon-KOON-doh)

Bowane (boh-WAH-nay): Lonkundo for civet cat

Embenga (em-BENG-ga): Lonkundo for pigeon

ika-o (ee-KAH-oh)

bwa-wa (BWAH-wah)

Nguma (uhn-GOO-mah): Lonkundo for python

swe-o (SWAY-oh)

Ulu (OO-loo): Lonkundo for tortoise

ta-ka (TAH-kah)

a-o (AH-oh)

ngo-nga (uhn-GOH-uhn-GAH)

pa-o (PAH-oh)

N-YEH (nyeh): Lonkundo for no or never, and traditionally spelled nye

Muh! (muh!): An expression of disgust, traditionally spelled mu!

One day in the town of Tondo, Bowane the civet cat met a beautiful feline he wanted for a wife. The cat was willing. And her father agreed to the marriage and set a certain bride price.

Bowane returned to his own animal village.
Soon he acquired the copper bars and
ornaments he needed. And early one morning,
with the bridewealth in a basket, he set out to
fetch his bride.
    Then there was
    Bowane walking, *ika-o, ika-o, ika-o*—
All alone, traveling to Tondo.

Bowane needed attendants to go with him. So he went to the home of his friend Embenga the pigeon and called, "Embenga, are you awake?"

Embenga peeked out of his small doorway. "Yes. I am awake."

Bowane said, "Come go with me to Tondo. There I shall marry a beautiful cat, and I need you to attend me."

"Between friends there is only goodness," said Embenga. "I will go with you."

Then there were

Bowane walking, *ika-o, ika-o, ika-o;*

And Embenga flapping, *bwa-wa, bwa-wa, bwa-wa—*

The two of them traveling to Tondo.

Next they went to the home of Nguma the python. Bowane called, "Nguma, are you there?"

Nguma poked his head out of his doorway. "Yes. I am here."

Bowane said, "Come go with us to Tondo. There I shall marry a beautiful cat. And I need you to attend me."

Nguma grumbled, "To Tondo! That's a day's journey! Why didn't you fall in love close by? But since you are my friend, I will go with you."

Then there were

Bowane walking, *ika-o, ika-o, ika-o;*

Embenga flapping, *bwa-wa, bwa-wa, bwa-wa;*

And Nguma slithering, *swe-o, swe-o, swe-o—*

The three of them traveling to Tondo.

Last they went to the home of Ulu the tortoise. Ulu was in his yard mending a fishing net.

Bowane said, "Ulu, come go with us to Tondo. There I shall marry a beautiful cat. And I need you to attend me."

"Oh, a wedding!" said Ulu. "I never miss a wedding. Of course, I will join you."

Then there were

Bowane walking, *ika-o, ika-o, ika-o;*
Embenga flapping, *bwa-wa, bwa-wa, bwa-wa;*
Nguma slithering, *swe-o, swe-o, swe-o;*
And Ulu waddling, *ta-ka, ta-ka, ta-ka, ta-ka*—
The four of them traveling to Tondo.

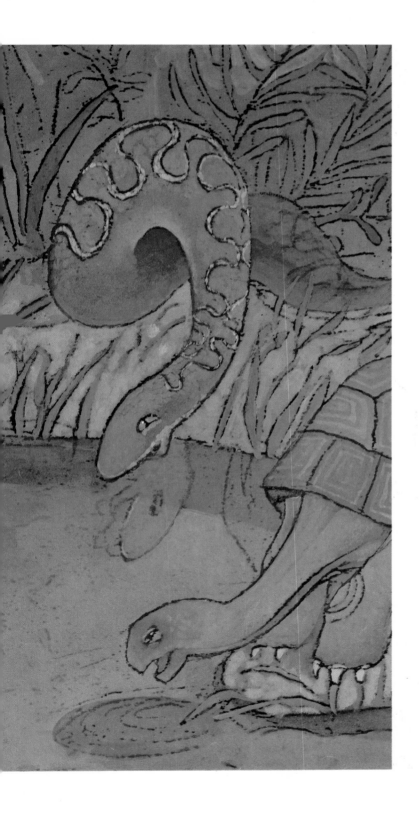

Presently the travelers came to a waterhole. The pigeon, the python, and the tortoise began to drink, *a-o, a-o, a-o.*

The civet cat was thirsty too. But he said, "It is taboo for me to drink water except from my own dish. Wait here. I will return home and fetch it."

His friends said, "No matter. That is all right."

So while they waited at the waterhole, Bowane went back home. After a long time he returned. "See, I have come already," he said. And he filled his dish and drank.

Then they went on—

Bowane walking, *ika-o, ika-o, ika-o;*
Embenga flapping, *bwa-wa, bwa-wa, bwa-wa;*
Nguma slithering, *swe-o, swe-o, swe-o;*
And Ulu waddling, *ta-ka, ta-ka, ta-ka, ta-ka—*
The four of them traveling to Tondo.

At length, they came to a palm tree that was loaded with nuts. Embenga was all aflutter, *bwa-wa, bwa-wa*. But the palm nuts were not ripe.

Embenga said, "Stay here, my friends, until the nuts are ready to eat. You know we pigeons like nothing better than ripe palm nuts."

"No matter," said Bowane. And the others said, "It is all right."

So they stayed near the palm tree for two whole weeks while the nuts ripened, and Embenga ate them.

Then they went on—

    Bowane walking, *ika-o, ika-o, ika-o;*

    Embenga flapping, *bwa-wa, bwa-wa, bwa-wa;*

    Nguma slithering, *swe-o, swe-o, swe-o;*

    And Ulu waddling, *ta-ka, ta-ka, ta-ka, ta-ka*—

The four of them traveling to Tondo.

Farther on, Nguma caught a small antelope. He squeezed it
and licked it. Then, A-OOOOOH! He swallowed it all in one piece.

And he said, "My friends, when I swallow an animal, I cannot travel until the digestion is finished. We must rest here."

"No matter," said Bowane. And the others said, "It is all right."

Day after day, the travelers waited and watched, watched and waited, as the lump in the python grew smaller. Finally it was gone. And Nguma said, "Arise. I can travel now."

And so they went on—

Bowane walking, *ika-o, ika-o, ika-o;*

Embenga flapping, *bwa-wa, bwa-wa, bwa-wa;*

Nguma slithering, *swe-o, swe-o, swe-o;*

And Ulu waddling, *ta-ka, ta-ka, ta-ka, ta-ka*—

The four of them traveling to Tondo.

At last they reached a forest that was near their destination. But there, blocking the path, was a huge fallen tree. The civet cat and the python climbed over it. The pigeon flew over it.

But the tortoise scrabbled up a little way and then fell back, *ngo-nga*. He did that again and again. Finally he gave up and said, "My friends, I am not able to climb over this big tree trunk. We must stay here until it rots, so that I can cross."

Bowane cried, "We cannot do that! When would we ever get to Tondo?"

Nguma echoed, "We cannot do that!"

And Embenga shook his small head and said, "N-YEH, N-YEH!"

But Ulu protested, "I waited for all of you! Why do you complain when you have to wait for me?"

So they stayed there in the forest year after
year, while the tree trunk rotted away. And one
day Ulu said, "At last! It is time to rejoice. The
log has crumbled."

And *pa-o, pa-o, pa-o*, he climbed over the
mound.

Then the friends continued their journey.
And soon they emerged from the forest and
entered the town of Tondo.

Straight to the home of the bride they went.

He called, "I, Bowane, have come!"

The beautiful cat Bowane had come to marry appeared at the door. "Muh!" she cried. "How dare you show your face after all these years? Did you think I would wait forever?"

Just then two civet kittens came tumbling out of the doorway beside her. "I married someone else," she said. "And these are my children."

"Oh," said Bowane sadly. Then he explained, "My friends waited for me to get my water dish. We waited for Embenga's nuts to ripen. We waited for Nguma's food to digest. And we waited for a log to rot so that Ulu could pass."

"You waited for a log to rot!" she cried. "How could you be so foolish! Now you and your friends be gone, or I'll call my husband."

Then out from behind her came the biggest civet cat Bowane and his friends had ever seen. And he was baring his teeth, *NNNNNNN!*

Suddenly there were
   Bowane running, *ikaoikaoikao;*
   Embenga flying, *bwawabwawabwawa;*
   Nguma streaking, *sweosweosweo;*
   And Ulu scurrying, *takatakatakataka—*
The four of them hurrying home from Tondo.
   As in this story, sometimes between
friends there is too much consenting. If a thing
is not wise to do, it is best to say, "N-YEH!"